I0559596

FROM A HEART SURGEON
TO A COOK

Dr. Luis Mispireta

Copyright © 2024 **Mispireta Family Foundation**

All rights reserved. No part of this publication may be reproduced, distributed, or transmitted in any form or by any means, including photocopying, recording, or other electronic or mechanical methods, without the prior written permission of the publisher, except in the case of brief quotations embodied in critical reviews and certain other noncommercial uses permitted by copyright law. For permission requests, write to the publisher, addressed "Attention: Book Rights and Permission," at the address below.

Published in the United States of America

ISBN 978-1-962569-47-7 (SC)
ISBN 978-1-962569-45-3 (HC)
ISBN 978-1-962569-46-0 (E-book)

Mispireta Family Foundation
222 West 6th Street
Suite 400, San Pedro, CA, 90731
www.stellarliterary.com

Order Information and Rights Permission:

Quantity sales. Special discounts might be available on quantity purchases by corporations, associations, and others. For details, contact the publisher at the address above.

For Book Rights Adaptation and other Rights Permission. Call us at toll-free 1-888-945-8513 or send us an email at admin@stellarliterary.com.

Table of Contents

Chapter 1

The Tantalizing World of Taste

My tastes are simple: I am easily satisfied with the best."
—Winston S. Churchill (Hale, 1966) or Oscar Wilde? (Saltus, 1917)

How do we enjoy our food?

The human senses do not work independently; often they work in synchrony to produce an experience that may or may not be pleasurable.

Our ability to distinguish noxious substances from nutrient-rich foods is essential for our survival. Although olfaction and vision participate in this function of food identification, the sense of taste is the ultimate checkpoint to differentiate nutrients from noxious elements.

With respect to food, eating is a continuum from the need to have an intake of food (calories, carbohydrates, proteins, and fats) to satisfy a biological need, to the experience of having a pleasurable dinner well prepared and presented, accompanied by beverages of your choice that match the dishes served, and perhaps soothing music.

Even though it is true that **satisfying a biological need is pleasurable** (if thirsty, a glass of water tastes so good), once that has been accomplished, the pleasurable experience of eating can be emphasized by other senses. These include those that are chemically mediated, which means a chemical compound stimulates the sensor (taste and smell), and other senses that are not chemically mediated, such as feel, sight, and hearing. We see a visual image of the food's presentation, hear the sizzling sound of the fajitas as they are brought to our table, and experience the feel in our mouths of heat, texture, and the sleekness or astringency of the preparation. Sleekness refers to smoothness, like a creamy sauce. Astringency refers to the sensation of dryness produced by certain foods or drinks like tannins in wine.

Taste is the experience perceived by the taste buds distributed in the mouth and transmitted to the brain centers that allow us to recognize tastes and assign them as being pleasurable or unpleasant. There are five recognized tastes: sweet, salty, sour, bitter, and most recently recognized, umami or savory. (Roper & Chaudhari, 2017)

Flavor is a broader sensory experience. It includes the different potential combinations of tastes and proportions all at once, including the input from all other senses, particularly from the aroma of the dish. Flavor will be the topic of Chapter 2.

Since ingredients contain multiple tastes in specific proportions, and a dish contains multiple ingredients, the product of them will produce thousands or multiples of thousands of variations and then, multiplied by the different methods of cooking, creates an infinite number of flavors.

And yes, you, the home cook, can create flavors particularly pleasant to you or your family, friends, or guests!

At times, these combinations that are created become unique, new, and identifiable, and different from the ingredients. Daniel Patterson and Mandy Aftel in their book The Art of Flavor (Patterson & Mandy, 2017) define it as **locking**, where the final sum is greater than the parts. They also describe another phenomenon where the combination of ingredients minimizes certain aspects of the taste of some of the other ingredients. They call this **burying,** using the balancing properties of the different tastes. One such example is the burying of the sweet taste by umami rich products (Melis & Barbarossa, 2017), or the bitter taste of coffee by the sweetness of sugar.

Gustation: sense of taste

Two senses use chemical sensors to start the process from stimulus to sensation. The sense of taste requires an immediate contact of the stimulus with the sensor (taste buds). The sense of smell, a remote chemical sense, does not require immediate contact with the stimulus (aroma) because the aromatic particles are carried in the air, so you could smell the aroma from a different room where the dish is being prepared. (Dunning, 2017)

Characteristics of tastes

For a flavor to be recognized as a sense of taste requires:

1. **A dominant molecular compound as the stimulus for each taste:**

- **Sweet = sucrose**
- **Salty = sodium chloride**
- **Sour = acid** (acetic acid, citric acid, etc.)
- **Bitter = alkaloids** (quinine, caffeine)
- **Umami** (savory) is rather complex and will need further elucidation, see below.

2. **A specific receptor for each taste. Sweet** substances bind to G-protein-coupled receptors (like bitter receptors), leading to nerve activation. **Bitter** receptors are also G-protein-coupled and detect alkaloid-based substances. The number of these receptors diminishes with age, and maybe that is why we stop using sugar in our coffee or tea as we grow older. **Sour** taste receptors have acid-sensing cation channels that detect H^+ ions, their concentration determines the degree of sourness. **Salty** taste receptors have Na^+ ions channels, and the concentration of Na^+ ions determine the saltiness of the stimulant. Because sodium is used in so many biological processes, the hormone aldosterone increases the number of sodium channels in taste cells when there is a deficiency of sodium.

   Children have more bitter receptors than adults and will lose some as they age, one may use the balancing effect of sweet, salty, or sour on the bitterness of vegetables to help them enjoy them. This may explain as well how our preference for tastes changes as we age.

3. **Tastes should be irreducible** (not able to be further reduced; unique), not a combination of other tastes.

Umami or savory: In 1908, Professor Ikeda discovered monosodium glutamate (**MSG**) in seaweed, later determining that it is present in all savory foods in different degrees. In 1913, Inosine Monophosphate (**IMP**) in bonito, (a fish) flakes and in 1960 Guanosine 5 Monophosphate (**GMP**) were added as substances that produce the umami taste.

In about 2000, a publication from the University of Miami reported the presence in the human tongue of a receptor that specifically responds to the glutamate ion, fulfilling the requirements for classifying umami as a sense of taste.

It seems that a synergy exists among them (MSG, IMP, and GMP) when used in a proportion of 1 to 1. IMP with glutamate, makes the umami activity more effective or more active by 7.5 times, and when GMP is used with glutamate in the same proportion, it makes the umami more effective or active by thirty times.

Additional tastes to consider:

- In addition to the five basic tastes, other sensations occur in the oral cavity as we eat. They contribute to the pleasurable experience of dining, and they include the sensation of **heat** or **cold** (which also modify the basic tastes), **texture of the ingredients**, in particular proteins and vegetables, and **astringency**, such as with tannic wines. These sensations are mediated through the sense of **feeling** or the **tactile** sense.

- **Spiciness** is a sharp, hot sensation produced by certain substances, such as Ajíes or chilies of the capsicum family. The chemical identified in this spice is capsaicin and the sensor for this substance is called the VRI receptor. Capsaicin in this manner fulfills two requirements; being irreducible would be the missing third requirement. There is a tremendous variety throughout the world, with many degrees of pungency, to the point that the Scofield scale was designed, which assigns a capsaicin value to each variety of chili pepper. Values vary in a broad spectrum from very mild to extra hot. The value given to each species varies from as few as one hundred units to over a million units. Some of the milder ones may have some savory secondary flavors that enhance the umami taste of some dishes, especially when used in conjunction with onions and garlic.

- Some other tastes to consider are metallic or fatty. At this stage there is not enough documentation for them to be added as tastes.

These tastes interact with each other, sometimes complementing or at other times balancing each other. This happens because they may use similar or the same receptors, or the same chemical compounds that make these sensors work. The reasons other interactions happen are still not known, but observation and general knowledge make the interactions familiar, as seen in the diagram below. Some of these may also explain why some pairings work so well, having become internationally known and used frequently, like tomatoes and basil, or tomatoes and cheese. These interactions have been represented graphically, and in the graph below spiciness has been added, even though is not recognized as a taste yet.

For the flavor of a new dish to be balanced should contain two or more of the tastes of the figure 1 below. They may be in the category of balancing each other or complementing each other.

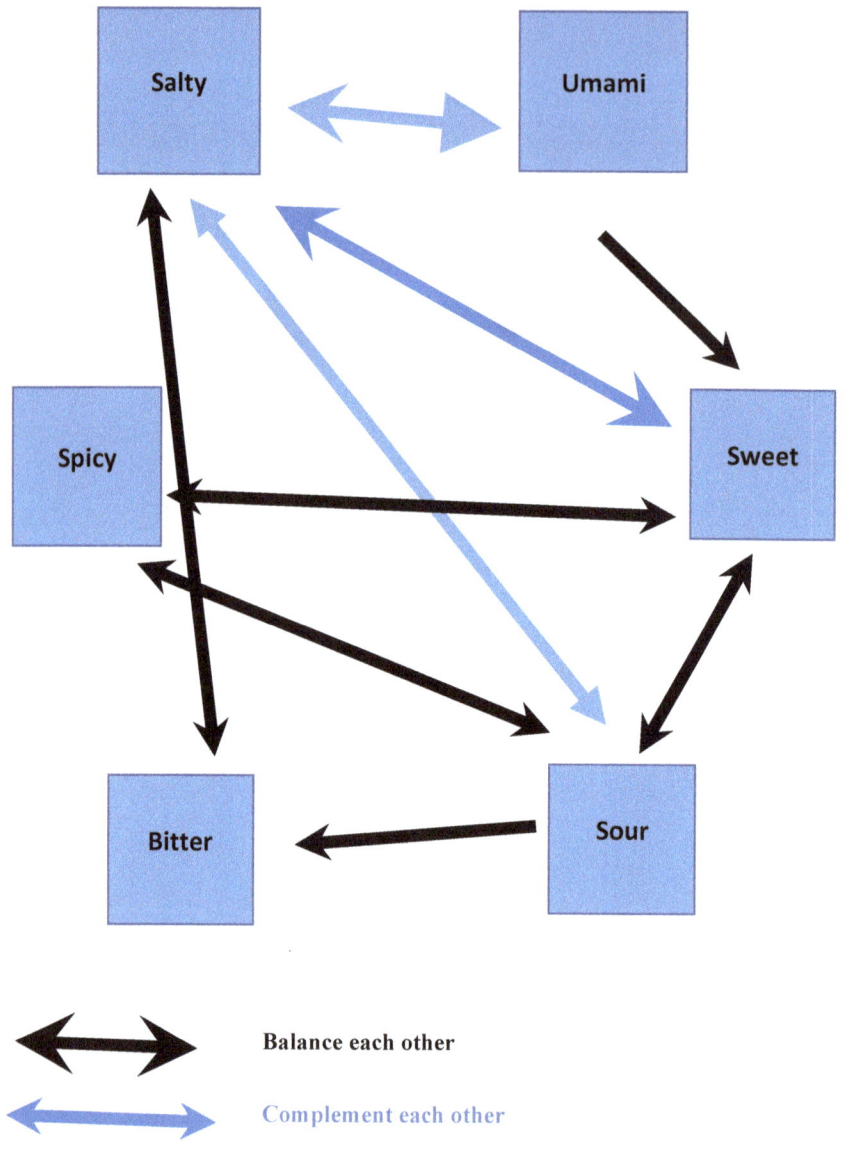

Salty ←→ **Umami**

Spicy

Sweet

Bitter ← **Sour**

←→ Balance each other

←→ Complement each other

Tastes and inter relationships

Chapter 2.

The Flavor Experience

"Variety's the very spice of life that gives it all its flavor."
(Cowper, 1913)

Flavor is a broader experience than taste and involves senses other than the taste sense, mostly the olfactory sense, but the visual, hearing, and feel senses can participate as well.

The flavor of a dish is the result of the combination of ingredients with multiple tastes in each one. An ingredient contains multiple chemical compounds, each with a distinct flavor. The flavor can be modified and multiplied further by the cooking methods chosen, or by the ripeness of the ingredient, timing of the harvest, the geographical location of the cultivar, etc. Examples include grapes for wine or tomatoes for a salad or sauce. There are an infinite number of flavors, and new ones are created daily.

Sometimes some dishes develop a recognizable flavor with different types of cuisine, such as Far East, Middle East, European, or South American. Some are even recognizable by regions of the same country, such as northern versus southern Italian. Because many of the ingredients are indigenous to microclimates, there can be significantly different recognizable flavors in neighboring villages. Despite having the same ingredient, grapes, for example, produce different flavors depending on the variety (Concord and Riesling, for example), ripeness, time of harvest, and more. (Brantley, 2018)

The flavor of an ingredient is an amalgam of multiple flavors, and as we use multiple ingredients, we may end up with hundreds of flavors in variable proportions, creating the possibility of an infinite number of dishes.

The only limit is the cook's creativity and willingness to experiment.

Classes of flavors

With such a great variety of flavors, there is a need for organization and communication in a common language.

Organization seems to be commonly achieved by similarity with musical notes: **high notes** (herbs and spices) **middle notes** (some spices, salads, poached chicken) and **base notes** (roasted meat or chicken and umami-laden dishes).

The language most used refers to a **shape** of the flavor: angular (sharp to blunt) or round (mouth filling, weighty). We also have **qualities**, such as fresh (ginger, coriander, lemongrass); sharp (black pepper, white pepper, juniper berry); sweet (anise, nutmeg, allspice, cinnamon, cardamom, clove, chervil, parsley, dill, tarragon, spearmint, basil, cilantro, and peppermint); citrus (lemon, lime, grapefruit, orange); earthy (saffron, turmeric, cumin); and savory (thyme, rosemary, oregano, and sage).

Rules of flavor

Since the main purpose of cuisine is to produce dishes that are pleasurable, and even though some combinations of ingredients are known to produce dishes where the result is greater than the sum of the ingredients, the search for the right one for the home cook can be exhausting.

In addition, the decision must be based on what is in the refrigerator and the pantry. So, the home cook may have the ingredients without a recipe and will have to create a dish with what is at hand. We will endeavor to present a way of accomplishing that.

To develop a system to create dishes from what we have at hand or to select ingredients we want to use, Daniel Patterson and Mandy Aftel (Patterson & Mandy, 2017) suggest some rules to orchestrate and facilitate this.

Rule 1 - Similar ingredients need a *contrasting* flavor in each ingredient, to differentiate them (e.g., potatoes or rice contrasted with chili sauce in the potatoes and cilantro, cucumber, or anise in the rice).

Rule 2 - Contrasting ingredients need a *unifying* flavor, bridging or linking opposite flavors (e.g., feta cheese and spicy ají amarillo bridged, or linked, with oil and cream).

Rule 2 Example:

Huancaína Sauce

Sauce based on farmer cheese, ají amarillo (contrasting flavors, Rule 2) unified by cream and oil.

Ingredients:

- ½ cup Ají amarillo paste
- 2 Tbsp. vegetable oil (canola or grape seed oil)
- ½ cup heavy cream or half and half
- ½ lb. farmers cheese soaked in milk to reduce saltiness
- Soda or water crackers (optional) cracked ¼ cup or less for thickening.

Procedure:

1. Place all ingredients in a food processor or blender.'
2. Process until smooth. Refrigerate until ready to use.

Can be used in a variety of ways, dipping sauce for French fries or fried yucca (cassava), pasta, risotto and the traditional way over boiled potatoes with Alfonso olives and slices of hardboiled egg.

Huancaina Potatoes

Another example could be Roasted beef with wasabi mashed potatoes and rice spiced with fennel seeds (2 carbs differentiated by the spices).

Rule 3 - Heavy flavors need a lifting note (e.g., beef with rich creamy mushroom sauce and blueberries, the blueberries doing the lifting).

Rule 4 - Light flavors need to be *grounded* (e.g., fat, earthy herbs, or spices, or fermented umami-rich ingredients like soy sauce, fish sauce, or anchovies) (Tower, 1986)

Rule 4 Example:

Bell Pepper Salad with Anchovies, Olive Oil, and hearts of palm

Light flavors grounded by anchovies and unified by olive oil)
Serves 6-8
Ingredients:

- 12 Bell peppers, mixed green, yellow and red
- 8 fillets of white Spanish Cantabria anchovy in olive oil.
- ½ c. EVOO (extra virgin olive oil)
- ¼ c. white wine vinegar
- Salt and pepper to taste (1/2 tsp. salt and ¼ tsp. pepper)
- 6-8 black olives (Alonso olives work best)
- Herbs to taste (basil, oregano, thyme, or sage)
- Five stalks of hearts of palm split lengthwise.

Procedure:

1. Grill or roast bell peppers as slow as possible, turn frequently. Place them in a pan and cover with aluminum foil to cool.
2. Open them when cool, scrape the seeds and ribs, trim tops and bottoms and divide in sixths and lay them in a platter.
3. Open the anchovies tin, put them in a plate cover them with fresh olive oil and rest for 30 minutes. If the taste of the anchovies is too strong, soak them in milk for 20-30 minutes, rinse them in cold water and then proceed with fresh olive oil.

4. Prepare the vinaigrette with the vinegar, salt pepper and ½ cup of EVOO, pour over peppers.
5. Arrange the olives and anchovies over the peppers, add the hearts of palm. Then Serve.
6. May place the olive oil from soaking the anchovies in small dishes for bread dipping.

Roasted peppers, hearts of palm and anchovies' salad

When pairing ingredients for a dish try to maintain a balance. To achieve this should incorporate three or four of the five tastes. Use the elements of the Figure 1 and follow the balancing arrows (black) to achieve the balance that tastes good to you. (Brantley, 2018)

Another example would be Stir fried cauliflower rice with scallions and soy sauce.

Modifiers of Flavors

The most potent accents of flavor are spices and herbs because of the compounds they carry. An herb or spice contains more than one compound, so they end up with a primary flavor and secondary flavors.

For this reason, we will give them significant attention. They will be arranged by primary flavor; their secondary flavors will be listed; and compatibility with other ingredients will be mentioned when appropriate.

SPICES

Sweet Spices

Anise Primary - sweet, sweeter than sucrose
Secondary - mint and licorice
Pairs with citrus fruits, pineapple, melon, coconuts, figs, carrots, coffee, goat cheese, shellfish, fish, lamb.
Pairs with other herbs/spices: mint.

Nutmeg Primary - sweet bitter
Secondary - clove
Pairs with cabbage, carrots, chicken, eggs, fish, lamb, onion, potatoes, sweet potato, pumpkin, spinach.
Pairs with other herbs/spices: cardamom, cinnamon, cloves, coriander, cumin, ginger, thyme.

Allspice Primary - like cinnamon, cloves, and nutmeg all together
Secondary - sweet balsamic and sweet flowers
Pairs with tomatoes, BBQ sauce-tomato based, all orange vegetables, good in savory dishes like grain salads and polenta. Allspice pairs well with beets, carrots, parsnips, and winter squash.

Cinnamon Primary - sharply aromatic, sweet, warm, and a bit bitter
Secondary - clove-like accent
Pairs with tomatoes, rice, oatmeal, chicken, fruits such as blueberries, apples, apricots, almonds, pears, and bananas.
Pairs with other herbs/spices; it does well with basil, becoming sweeter but simultaneously spicier.

Cardamom Primary - sweet, bitter
Secondary - mint, eucalyptus, and pine medium intensity
Pairs with sweet and savory dishes, fruits, legumes, sweet potatoes, and other root vegetables. Good in coffee, tea, and rice dishes.
Pairs with other herbs/spices: caraway, cinnamon, cloves, coriander, cumin, and ginger.

Clove   Primary - sweet sharp fruity and woody
Secondary - peppery camphor and floral
Pairs with beets, red cabbage, ham, pork, pumpkin, squashes, sweet potatoes, and apples.

Sharp Spices

Black Pepper Primary - woody, clean, penetrating, sharp
Secondary - clove and lemon facets with additional floral hint
Used universally in dishes including fruits, salads, and all meats, providing a counterpoint to meaty, creamy mushroom sauce.

White Pepper Primary - like black pepper but more pungent

Green Pepper Mildest of all peppers

Juniper Berry Primary - bittersweet
Secondary - pine and resin, woody and astringent like gin
Pairs well with lamb (or mutton) and is particularly good with venison, wild boar, and even domestic pork. You could even add the berries to a pot of chili to give a rustic flavor that complements ground smoked chili peppers. Juniper is also a good flavoring to use with roast duck. Pairs with other herbs/spices: garlic, lime, lemongrass, cilantro, basil, mint, scallions, turmeric.

Fresh Spices

Fresh, bright character in the middle to top notes. Airy, provide a lift.

Ginger Primary - fresh and light, sweet and warm when fresh. When cooked, the aroma diminishes but pungency remains or increases.
Pairs with chicken, fish, apples, passion fruit, pears, pineapple, and mango.

Pairs with other herbs/spices: garlic, lime, lemongrass, cilantro, basil, mint, scallions, turmeric.

Lemongrass Primary - sour, clean, and lemony
Secondary - rosiness
Pairs with fish, chicken, pork, coconut, coconut milk soups and stews.
Pairs with herbs/spices: basil, cilantro, cinnamon, cloves, garlic, ginger, and turmeric.

Coriander Seeds Primary - sweet floral
Secondary - peppery
Pairs with apples, beef, chicken, eggs, ham, pork, lentils, onions, plums, and potatoes.
Pairs with herbs/spices: cinnamon, cloves, cumin, garlic, ginger, fennel, and nutmeg.

Earthy Spices:

They have a dampening, flattening effect, down-drawing to lower notes and richer meaty flavors, work harmoniously with counterpoint fresh bright flavors like citrus and sweet spices.

Saffron: Primary - rich, delicate, lingering, difficult to nail down its character – floral? Honey Pungent? But easy to recognize.
Secondary - musky, sweaty, bitter
Pair's best infused with liquids in soups or water/broths for rice cooking.

Cumin Primary - bittersweet, heavy and dense, nutty, with multiple facets; some hints of orange and ginger. Has an intense sweaty, pungent aroma.
Pairs with beef, beans, chickpeas, couscous, eggplant, lentils, potatoes, rice, squash, and tomatoes.

Turmeric Primary - bitter, sour, earthy, color more yellow than orange
Secondary - pepper and ginger

Pairs with fish, meat and poultry, eggs, beans, lentils, rice, root vegetables, spinach.

Pairs with herbs/spices: cilantro, cloves, coriander, cumin, fennel, garlic, ginger, and lemongrass.

Herbs

The leaves and flowers of the plants are more delicate than spices (barks, stems, and seeds) and therefore should be added at the end of the cooking process.

Sweet herbs

Bright and light, their function is to introduce the different flavors of the dish. They are connectors, bridging disparate flavors, like citrus and floral flavors do.

Chervil   Primary - light sweetness and aroma like anise, flavor like a mix of tarragon and parsley

Secondary - caraway, pepper, and parsley

Pairs with beans, cheeses, chicken, cream, eggs, fish and seafood, turkey, arugula, asparagus, beets, broccoli, rabe, carrots, celery, endive, lettuce, green beans, kale, mushrooms, onions, peas, potatoes, shallots, spinach, tomatoes, watercress, zucchini.

Pairs with herbs/spices: basil, chives, dill, lemon thyme, mint, mustard, parsley, and tarragon.

Parsley   Mostly flat-leaf parsley, others almost no flavor

Primary - sweet, tangy, grassy, herbaceous

Secondary - lemony note

Freshens a dish without imposing a strong presence.

Pair with eggs, fish, most vegetables, tomatoes, lentils, and rice.

Pairs with other herbs/spices: other sweet herbs, chives, garlic, mint, oregano, rosemary.

Dill Primary - mild sweet, clean
Secondary - citrus notes
Lifts delicate ingredients without overpowering them.
Pairs with fish, chicken, avocados, beets, cabbage, carrots, celery, cucumber, potatoes, rice, tomatoes, yogurt, zucchini.
Pairs with herbs/spices: basil, parsley, garlic, ginger, turmeric.

Tarragon Primary - sweet and green
Secondary - anise and basil facets
Pairs with fish, poultry, eggs, artichokes, potatoes, asparagus, tomatoes, zucchini.
Pairs with other herbs/spices: basil, chives, dill, parsley, basil (other sweet herbs), fennel and anise seeds.

Spearmint Primary - sharp pointed character, cooling quality but not as pronounced as peppermint.
Secondary - fruity facets of mango and pear and a lemony touch

Peppermint Primary - powerful, easily recognizable, has a bite and strong cooling sensation and aroma.
Secondary - peppery and jasmine facets

Basil Primary - sweet and spicy character, green and herbaceous quality, and lightly peppery
Secondary - clove and anise, many types of basil, sweet basil most used in the western world.
Pairs well with tomatoes, bell peppers, peas, potatoes, rice, corn, white beans, zucchini, and fruits like apricots, blueberries, peaches, tomatoes and soft cheese.
It pairs with other herbs/spices: chives, cilantro, mint, parsley, rosemary, and thyme.

Cilantro Primary - sweet spice and piney
Secondary - lemon and anise facets, some peppery hint
Pairs with soups and stews, root vegetables, potatoes, carrots, cucumber, rice, corn, coconut milk, avocado, figs, and yogurt.

Pair with other herbs/spices: garlic, ginger, lemongrass, mint, parsley, basil, chives, and dill.

Savory Herbs (Resinous)

In contrast with other herbs, these are sturdy and need to be cooked to mellow them out, like spices. They retain their flavor and power when dried.

Rosemary Primary - camphor, eucalyptus, pine, somewhat bitter and woody
Secondary - pepper, clove, and sage tones
Pairs with chicken, eggs, pork, fish, soups, stews, beans, lentils, eggplant, bell peppers, tomatoes, potatoes, and stuffing for poultry.

Thyme Most potent fresh rather than dry; loses the aroma with heat.
Primary - sweet, warm, herbaceous with biting quality but not bitter
Secondary - camphor and clove notes
Pairs with chicken, lamb, meats, fish, soups and stews, beans, eggplant, cabbage, carrots, corn, potatoes, tomatoes, winter vegetables.
It pairs with other herbs and vegetables: garlic, basil, lavender, nutmeg, oregano, parsley, rosemary.

Oregano Most potent dry rather than fresh.
Primary - savory, sharp, herbaceous, bitter, and warm.
Pairs with chicken, eggs, fish, pork, lamb, meats, anchovies, beans, eggplant, cabbage, cauliflower, corn, potatoes, sweet peppers, squash, tomatoes, zucchini, and pizza.
It pairs with other herbs/spices: garlic, cumin, basil, parsley, rosemary, sage, and thyme.

Sage Primary - sweet, bitter, sour, and savory (all tastes but salty), herbaceous, musky
Secondary - eucalyptus bittersweet after taste

Pairs with chicken, oily fish, fatty meats, goose, liver, soups and stews, beans, stuffing, asparagus, potatoes, and tomatoes. A way to preserve fresh herbs is by preparing pastes. Here is a recipe for preparing cilantro paste. You may replace cilantro by any other herb of your preference.

Example recipe: Cilantro Paste

Ingredients:

- 3 cup cilantro leaves washed and dried, lightly packed. Or herb of your preference
- 1 tsp minced garlic
- ½ cup evaporated milk or coconut milk
- ½ cup of roasted unsalted cashews
- 2 Tbsp. of lime juice
- Salt to taste

Procedure:

1. Mix all ingredients in a food processor and puree to a fine paste.
2. Place the paste in an ice cube tray and freeze.

Cilantro Paste frozen.

Other modifiers of flavors:

Onions: Not only an Ingredient but they also modify the flavor of dishes, depending on how they are used. There are many types of onions red onions are more piquant than other types; white onions are milder than red onions; sweet white onions like Vidalia onions, shallots, etc. The way to cook them affects their flavor as well.

When raw, they have an aggressive sharp, pungent flavor, which can be tamed by placing them in a sieve, adding a large amount of salt, rubbing the salt in the onions with your hands for a minute or so, then rinsing them with hot water, making sure to remove most or all the salt. The result of this exercise will be sweeter, less aggressive, and not-so-sharp onions due to neutralizing a chemical (sulfenic acid [R-SOH] liberated when the onion is cut) of the onion with the sodium of the salt. Then you can use the treated onions raw for salads or dressings with oil and vinegar and a bit of ají, perhaps for lentils or bean salads.

Marinated, still piquant but not as aggressively sharp.

Cooked, they mellow, are sweet, not sharp anymore but develop a round and expansive character. Best sautéed on low heat.

Charred, smoky and sweet.

When cooked with garlic or/and ginger, ají form the foundation of many dishes in the Iberia Peninsula, Central and South America and the Caribe, known as **soffrito or aderezo**. In the Creole and Cajun cuisine, they are paired with sweet peppers and celery and known as the holy trinity (ratio 1:1:1) and if carrots replace the sweet peppers, known as mirepoix (2 parts onion, 1 part celery, and 1-part carrots).

So, you can see the many personalities an onion takes, depending on the type of onion and the cooking method.

Example recipes: Vidalia Onions Appetizer

Vidalia onions roasted or microwaved as an appetizer and marinated raw onions in South American creole sauce.

INGREDIENTS

4 Vidalia onions (medium - large)

4 Tbsp. butter

4 Beef bouillon cubes or equivalent amount of granular bouillon.

INSTRUCTIONS

1. Slice the Top end of each onion and peel the outer layer.
2. Slice each onion into eights without cutting the whole way through. Leave them attached about 1/2 inch at the bottom, root end. Easer to place the onion in a ramekin of the appropriate size for the onion, were the edge of the ramekin stops the knife. Place in a Pyrex dish with the ramekins where they fit snugly but not touching each other.
3. Microwave on high for 7 minutes per onion.
4. Remove from microwave, place a tsp. of granulated bouillon down into the center, olive oil and balsamic vinegar. May remove the ramekin at this stage to allow the onion to open blooming.
5. Place back in the microwave in high and cook for 5 additional minutes per onion. Serve by placing in a dish. The onion will open, blooming.

You can melt a slice of cheese over the top of these to give them more of a French onion soup flavor.

Vidalia onion cut in eights.

Onion creole sauce (pickled onions)

1. **Tame the onions**: In a strainer, salt the sliced onions abundantly, massage them with your hands for a minute or so, then rinse with hot water to remove most or all the salt. Place them in a bowl.

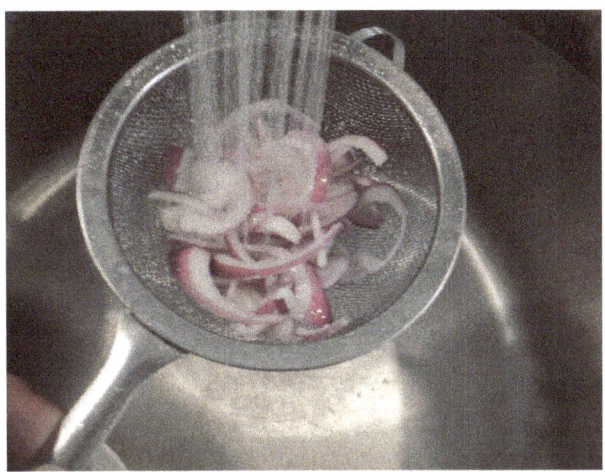

After salting them, rinse with hot water

2. Combine the washed onions with the sliced seedless ají amarillo or the paste, as well as   with salt, pepper, lime juice, and olive oil. Mix gently and sprinkle with the cilantro leaves.

Creole Sauce

Anchovies: Increase the savory taste (umami) of many dishes, e.g., Caesar salad dressing, or mushroom creamy sauce to accompany grilled steak. Example recipe: Mushroom creamy sauce.

Steak Umami Rich sauce: Prepared by sauteing shitake mushrooms, adding veal demi-glaze, and thickening by reduction.

Caviar: Increases the savory taste.

Truffles: Increase the savory taste.

Chili pepper: Piquant and pungent, but when mild, increases the savory taste (e.g., ají amarillo, guindilla). Ají amarillo can be found in multiple recipes that will follow, is easily found in the US at Latin American grocery stores or through the internet.

Combination of herbs and spices: Herbs and spices are used in certain combinations in different geographic locations, to the point they become recognizable for each location, for example:

> Italian: garlic, oregano, and basil
> Central and South American: cumin, coriander (seeds), cilantro (leaves), garlic, and ají
> French: marjoram, thyme, rosemary
> Chinese: anise, clove, cinnamon, peppers, fennel seed
> Indian: turmeric, cumin, coriander, red pepper, curry
> Spanish: pimenton (smoked paprika), garlic, saffron, cayenne.

Chapter 3

A Palette for Every Palate

"No one is born a great cook. One learns by doing."
—Julia Child (Child, 2018)

The painter has a wide range of colors to choose from on his palette; and similarly, the cook has a vast plethora of ingredients to choose from when creating art in the kitchen.

In this chapter we are going to look at the mix of foods needed to provide the calories and the building blocks for our bodies.

These include carbohydrates, proteins, fats, and water. They are macronutrients. In addition, there are several other nutrients that are ingested in small amounts called micronutrients, like minerals, antioxidants, phytochemicals, etc. Often, they come in ingredients that contain the macronutrients.

These nutrients can exist or present themselves in different states: solids, liquids, and gases. The state depends on the kinetic energy of its particles (atoms, molecules) and on weak bonds that keep them together.

Applying heat to these different states increase the motion of the particles, overcoming the weak bonds and allowing changes of state. The temperature at which this occurs depends on the nature of the substance. The temperature needed to change from solid to liquid is the melting point, e.g., the melting temperature for butter is 91°-95°F (32°-35°C). The temperature for changing from liquid to gas is the boiling point, which for water is 212°F (100°C).

Some substances do not go through these phases; carbohydrates and proteins will not melt nor vaporize, and sugar will melt but not vaporize.

Macronutrients

These are the main components of the human body. They provide us with calories (energy) necessary for normal function. On addition there are other particularly important functions provided by the macronutrients, some are somewhat obscure like cellular structure trough cell membranes (fats) including coverage of cell of the nervous system (myelin); Mitochondrial function (Proteins) lungs of the cell.

The Macronutrients are Carbohydrates, Proteins, Fats and Water. They come from the Plant Kingdom and the Animal Kingdom and Water from rain, rivers lakes and now in certain regions from the sea (desalination).

In general, the macronutrients except for water are synthetized from relatively simple compounds formed by carbon, hydrogen, oxygen, and nitrogen. These simple compounds will combine among themselves to form more complex molecules. While doing these will liberate water for each joining bond. This reaction is called Condensation. This is the process of synthesis.

Through cooking and digestion, the complex compounds will break down to simple compounds (opposite than synthesis) and will need water to complete their simple molecules. This process is called Hydrolysis.

The simple compounds are called Mono or disaccharides for Carbohydrates, amino acids for Proteins and fatty acids and alcohol compounds for fats.

A complex carbohydrate may have thousands of monosaccharide molecules, similarly proteins will have thousands of amino acids and fats with triglyceride, being the prevalent compound of oils, combines glycerol (3 carbon with 3 alcohol radicals) with three fatty acids.

These complex molecules become our food because they are palatable and trough time, we have learned to improve their taste by cooking.

Effects of cooking: the heat of cooking makes some changes in the macronutrients that makes them more palatable and facilitates absorption by our digestive system.

- Carbohydrates will break down into smaller compounds like dextrin's (crust of bread) and amylose (at 150-220°F (65.5-105°C). Another reaction carbohydrate will undergo with heat is Caramelization which is a complex reaction with production of caramels at 230-320°F (110-160°.

- Proteins with increasing temperatures will coagulate and agglutinate, bringing the protein to shrink in volume and lose weight. The typical flavor of browned meat comes from the Maillard reactions, which is a non-enzymatic complex reaction of amino acids and reducing sugars, it occurs at 280-330°F (140-165°C). Notice that is higher than the recommended internal temperature, so to obtain the Maillard effect need to sear the protein at high temperatures so the surface will sear without overcooking the inner portions.

- The solid fats with heat will liquefy, the oils will heat up, smoke and boil but will not evaporate. The greasy taste of fats can be modified by emulsification where small globules of fat are suspended in the other liquid, e.g., water, vinegar etc. There are two types of emulsions:
    1. Fat globules suspended in watery fluid (Mayonnaises)
    2. Watery globules suspended in fat. (salad dressing)

These explain why mayonnaise does not taste oily, but salad dressings do.

Smoking points of some fats and oils.

Depending on the temperature at which you want to sauté or fry the food product you are preparing, you may choose an oil with a smoking point above the temperature required for your dish. (Moncel, 2019).

Table 1 Smoking Point Temperature of fats

Fat Type	Smoking point °F	Smoking point °C
Butter	350°F	175°C
Butter, clarified	450°F	230°C
Lard	370°F	185°C
Vegetable shortening	360°F	180°C
Olive oil, light/refined	465°F	240°C
Peanut oil	450°F	230°C
Soybean oil	450°F	230°C
Corn oil	450°F	230°C
Canola oil	400°F	205°C
Avocado oil	400°F	205°C
Sesame oil	380°F	193°C
Coconut oil	350°F	175°C
Grape seed oil	450°F	230°C

Recommendations on consumption of Macronutrients

The total caloric intake recommended is of 1600 to 2000 calories per day for women and 2000 to 3000 calories per day for men. The lower end is for sedentary individuals and the high end for regularly active individuals. Most people are probably somewhere near the middle.

(US Department of Health and Human Services & US Department of Agriculture, 2015)

Recommendation for protein intake

Is of 0.8 g. per day per Kg. of ideal body weight. So, for 75kg. lean weight men would be times 4.5cal per gram of protein would be 270 Cal. from proteins.

Recommendations for the consumption of fats

Avoid Trans-fats by checking the labels on packaged foods, looking for trans' fats content or in the ingredients list for "partially hydrogenated."

Avoid overheating fats and oils beyond their smoking point or reheating them to avoid the formation of Trans-fats or having the food product acquire an unpleasant taste.

Read more about the effect of heating/reheating oils here: Effect of heating/reheating of fats/oils, as used by Asian Indians, on *Tran's* fatty acid formation.[1] (Swati & Santosh, 2016)

Limit saturated fats to less than 10 percent of calories per day for a 2,000-calorie diet. 200 calories would be the maximum allowed per day or 12 gm (a little less than 1 Tbsp.) of saturated fat.

1. **Replace saturated fats with monounsaturated fats or polyunsaturated fats.** Use oils instead of solid fats and sauté with olive or avocado oil instead of butter. Remember that the caloric values of oil or butter are the same.

2. **Increase consumption of omega 3 polyunsaturated fats** primarily by eating more fish, maybe twice a week.

3. **Eat skinless poultry** since the saturated fats are in the skin primarily. Trim visible fat on beef and poultry. For beef, choose lean meats and

---

[1] SwatiBhardwaj Santosh JainPassi Anoop MisraKamal K.Pant KhalidAnwar R.M.Pandey VikasKardam, accessed June 21, 2016, https://doi.org/10.1016/j.foodchem.2016.06.021

grass-fed, which seem to have a higher content of Omega 3 polyunsaturated fats than conventional grain-fed cattle.

Do not go to extremes. Do not eliminate fats completely; they are still necessary since essential fatty acids cannot be synthetized by our bodies, and you could end up eating more processed food advertised as being fat free and eating too many carbs and sugars.

So, in a resume a 75 Kg. male medium exercise should have total caloric intake of 2500 calories, 270 from proteins, 20-30% from fats so 500- 750 calories. These should be divided 50-75 cal. maximum from saturated fats (butter ≈1 tbsp.) the remaining fat calories or so from monounsaturated, or omega3 polyunsaturated fats (fish) 450-675 kcal. (MayoClinic, 2021)

Recommendations for consumption of carbohydrates

The total intake of carbohydrates will be the difference between total caloric need minus the caloric value of the fat and protein intake. Let's say for 75 Kg. man, who has a caloric need of 2000 Kcal. And takes 270 calories in proteins (60 gm of proteins,), 20%-30% from fats or 400 to 700 hundred Kcal per day or 44 to 77 gr. of fat. The remaining calories must come from Carbohydrates or 2000 -175 and -450 = 845 Kcal. This would be then 188 gm. Of Carbohydrates. Best if they are complex carbohydrates, especially if they contain fiber soluble or insoluble, roughage.

This has been a simplified way of looking at the macronutrients, for the more curious the following subchapter may answer some of your questions.

Good source for references for recommendations on macronutrients is the Dietary Guidelines for Americans 2015-2020

SUB-CHAPTER 3 Expanded explanation of the Macronutrients

Carbohydrates

Carbohydrates are organic chemical compounds containing carbon, hydrogen, and oxygen. The origin can be from plants or animals. The building blocks of carbohydrates are the monosaccharides that are compounds of carbon, hydrogen, and oxygen with a simplified formula of $C_6H_{12}O_6$ with 5

OH^+ groups and one either a CHO^- or $CO^=$ group. The monosaccharides are glucose, fructose, and galactose.

When two of them combine, they form a disaccharide: sucrose, lactose, or maltose, depending on which monosaccharides are joining. When they join, they liberate a molecule of water, and this reaction is called condensation. To break apart, a disaccharide needs a molecule of water, and this reaction is called hydrolysis. This is important in the absorption and metabolism of carbohydrates. The disaccharide's simplified formula is $2(C_6H_{10}O_5)$ since they lose a molecule of H_2O when they combine.

Both monosaccharides and disaccharides are soluble in water, have a sweet taste, and counter, sour-tasting ingredients.

When starches break down to a 20-40 monosaccharides compound (most often when baking), they form a compound called dextrin that is still water soluble and sweet but to a lesser degree than mono- and disaccharides. An example is the crust of bread.

Starches include more complex carbohydrates, where hundreds to hundreds of thousands of monosaccharides combine, and cellulose (fiber), where several thousands of monosaccharides combine in a linear chain (not branched), forming fibers that provide rigidity to plant cells. They are generally not water soluble or sweet, and humans lack the enzymes necessary to digest fiber (cellulose).

Fiber can be of two types, insoluble or soluble.

Fiber type: insoluble (roughage)

Benefits: Regularity (relieves constipation), lower risk of diverticulosis (gut pouches that could get inflamed or bleed)

Food sources: Bran from grains/cereals, skins and seeds from fruits and vegetables, dried beans/peas, brown rice

Fiber type soluble

Benefits: Helps reduce straining with excretion, binds cholesterol in the gut, and helps control the rise of blood glucose after a meal by slowing down absorption.

Food sources: Fleshy part of fruits and vegetables, oats, dried beans, and peas.

Digestion and metabolism of carbohydrates

Ingested carbohydrates to be absorbed need to be broken down into monosaccharides (glucose, fructose, and galactose). This is achieved through mechanical means (chewing) and chemical means through catalyzers called enzymes. For complex carbohydrates they are called amylases.

Amylases are any member of a class of enzymes that catalyze the hydrolysis of starch (splitting of a compound by addition of a water molecule) into smaller carbohydrate molecules (disaccharides) such as maltose (a molecule composed of two glucose molecules). Further breakdown is accomplished by other classes of enzymes (disaccharidases) that break the disaccharides into monosaccharides (glucose, fructose, and galactose) that can be readily absorbed in the small intestine and transported by the bloodstream to the liver (chemical plant of the body) for different metabolic functions (such as energy production or storage).

The process of producing energy from the foods absorbed, metabolized through the metabolic pathways, comes from breaking down the linkages of monosaccharides (containing energy) to form smaller compounds. This energy is transported to where it is necessary (for all physiologic processes) by compounds that contain phosphorus like ATP. Carbohydrates produce 4.5 K calories per gram.

Cooking carbohydrates

In the kitchen with heat in the presence of water, changes occur, known as gelatinization. First the granules of starch swell by absorbing water. With continued heat, a second step is penetration of double helicoidally structures of amylopectin, and then a third step occurs when the disintegration of the granule structure, then amylase leaches out into the water.

The process of gelatinization of carbohydrates requires energy (heat) and water to break down the complex polysaccharides into smaller compounds (dextrin, amylose) since a molecule of water must be added for each linkage broken.

Multiple variables, like the nature of the starch origin (potatoes, sweet potatoes, corn, etc.), the acidity of the liquid, and the concentration of salt, sugar, fats, and proteins in the recipe influence the temperature at which gelatinization will occur. The gelatinization temperature will be between 150°F (65.5°C) to 221°F (105°C).

There is an inverse correlation between gelatinization temperature and glycemic index (how much blood sugar will rise after ingestion of a known amount of starch or sugar). The higher the gelatinization temperature of a specific starch or sugar, the lower its glycemic index. Gelatinization is used frequently in the kitchen to make starches more palatable or digestible or to thicken sauces.

Effect of cooking in proteins

Caramelization

This is a non-enzymatic browning and a very complex process that includes many chemical changes, resulting in a sweet, nutty flavor and a brown color. These chemical changes involve the creation of polymers, Caramelans ($C_{24}H_{36}O_{18}$), Caramelens ($C_{36}H_{50}O_{25}$) and Caramelins ($C_{125}H_{188}O_{80}$) that provide the color of caramel. They also involve the creation of volatile compounds like diacetyl (two CHO^- or $CO^=$ radicals) responsible for the characteristic flavor of caramel.

Since this happens at a temperature higher than that needed for water to boil, it also explains why foods only brown if prepared with dry heat methods, and why it only occurs on the surface since the water in the center of the food (fleshy parts of meats, vegetables, and fruits) does not evaporate, maintaining the center temperature at no more than 212°F. This also explains why caramelization of onions needs to be done slowly to have it happen in the interior of the food product.

These processes are used in the kitchen for making caramel sauce, dulce de leche, and candies, or to caramelize onions, potatoes, pears, and bananas. In general, they need to be done slowly, with moderate heat, to be able to control the process.

Since there are several sugars, monosaccharides, and disaccharides, each will caramelize at a different temperature. Fruit sugars (fructose) caramelize at 230° (110°C), maltose at 356°F (180°C), and all other sugars caramelize at 320°F (160°C).

Proteins

Proteins are macronutrients formed by long chains of amino acids (building blocks of proteins and our bodies). The amino acids have an amino group NH^3 and an acid group COOH joined to a radical R (backbone of the amino acids) that is a chain of carbons and hydrogen. What makes an amino acid different from the next is the radical R. There are only 20 amino acids but the sequencing of them is different for each protein and in the body, there are hundreds of thousands to millions of different proteins. Each one has a different function, such as producing energy (4.5 Kcal per gram), formation on enzymes (catalysts of biochemical reactions), hormones like insulin and many more, formation of structural proteins like skeletal muscles, bone, ligaments, teeth, blood vessels, and specialized muscles like the heart that, while healthy, do not fatigue. In addition are the immuno-proteins that defend our bodies from invasive organisms, transport proteins like hemoglobin (transports O_2) or lipoproteins (transport and remove cholesterol), and neurotransmitters essential for brain and nerve functions.

So, as we can see, proteins are essential to sustain life as we know it. In addition, a tissue that is rejuvenated daily is the skin. In a couple of weeks, practically all our skin has been changed. Similarly, the cover of the intestines is shed daily and must be replaced. All these processes are done with amino acids.

There are nine amino acids that we cannot synthetize, and for that reason are called essential amino acids. They are present in animal proteins, and some are present in different vegetables. In vegetarian or vegan diets, it is important that a mix of them is eaten daily to ensure that all nine essential amino acids are present. The remaining amino acids can be synthetized by the liver and are known as nonessential.

Absorption of proteins

The requirement for protein ingestion is about 0.8 gm/kg of lean weight, so the total daily requirement for a 70 kg person would be 56 g. The average adult eats about 80-120 grams of protein daily, more than the required amount. Once in the digestive system, through the action of enzymes called proteases the proteins are broken down to amino acids.

Once the amino acids are absorbed in the small intestine, they may follow many different pathways depending on the function for which they are destined. If they go to produce energy, they will be broken down to smaller compounds (maybe of 3 carbons) so they can enter the aerobic metabolic cycle (Krebs cycle) and produce the most energy. They may synthetize other proteins that produce structural proteins, to replace damaged cells like skin, intestinal mucosa, and others like immuno-proteins, hormones, or are used in tissue rejuvenation (apoptosis), etc.

Cooking effect in proteins

In the kitchen, when heat is applied, a process of coagulation or agglutination will occur, most often changing from a semiliquid state (colloid) to a solid state, such as frying or boiling an egg. This is a two-step process. First denaturation will occur, as when carbohydrates break down the complex compounds and take water to break down the links that held together the constitutions of the complex molecules (hydrolysis). If no water is present, the internal water will be consumed, and the protein will dry out. That is why it is more difficult to cook meat using a dry heat method than to, let us say, simmer or poach it with some liquid. It is easier to ruin a good piece of fish by sautéing to an overcooked status, than by simmering or poaching it in a mix of wine and broth.

With the continued rising of the internal temperature, the proteins will coagulate, or agglutinate and the fish will be done (internal temperature 145°F (62-63°C).

Table 2 Recommended internal temperatures for meats.

by the U.S. Department of Agriculture (USDA)

Product	Minimum internal temperature and rest time
Beef, pork, veal, lamb Steaks, chops, roasts	*145°F (62.8°C) rest for 3 minutes +*
Ground meats	160°F (71.1°C)
Ham, fresh or smoked (uncooked)	145°F (62.8°C) rest for 3 minutes +
Fully cooked ham (to reheat)	Packaged @USDA inspected plants 140°F (60°C); all others 165°F (73.9°C)
Poultry - whole bird, breasts, thighs, wings, ground poultry, stuffing	165°F (73.9°C)
Eggs	160°F (71.1°C)
Fish and shellfish	145°F (62.8°C)
Leftovers	165°F (73.9°C)
Casseroles	165°F (73.9°C)

Use an instant-read thermometer to check for doneness; any other way is guessing. Professional chefs may have other ways, but they repeat the same process many times a day, while we home cooks may repeat the same process a couple of times in a week.

Maillard Reaction

When cooking meats, part of the flavor comes from achieving a seared surface (golden brown color). It is achieved by getting the surface of the product at higher temperatures than the recommended internal temperature. This color and taste are obtained because of the Maillard reactions.

The Maillard reactions are multiple chemical reactions between amino acids and reducing sugars that gives browned food its distinctive flavor. It is a non-enzymatic browning that occurs rapidly at 280-330°F (140-165°C). If the temperatures get much higher, caramelization of sugars will occur and at the next level it will be burned.

This reaction occurs in the preparation of steaks, fish, bread, cookies, and deep-fried foods.

Fats

Fats or lipids are macronutrients used by our bodies as a fuel source and for storage, as well as aiding in the absorption of fat-soluble vitamins. They form protoplasmic membranes in all cells regulating the movement of all substances in and out of the cells. In the nervous system, lipids form myelin, a lipid that covers the axons, forming part of the spinal cord, the white substance of the central nervous system and the peripheral nerves. They play pivotal roles in the digestive system and in the hormonal system where a fat deficiency will produce a drop in hormone production.

As you can see, fats are essential for life as we know it, and total avoidance of lipids in our diet is not possible. Some of the recommendations in the past to avoid the intake of lipids resulted in substituting the fats for sugars and carbohydrates, facilitated by the processed food industry, resulting in a worsening of public health instead of an improvement.

Dr. Jose Revuelta, in an article of public dissemination of science about cholesterol[2], makes an insightful comment regarding the mechanisms of selection of food products. In the nomadic days of humans, hunters would leave their temporary habitats to go hunting for meat rich in fats. Even today when we are surrounded by food products, where hunting is not necessary for most of us, we still do not talk about how good the lettuce we ate last night was but will refer with fondness to the steak that came with the lettuce. This probably represents some genetic imprint that our ancestors had, that is still present today.

Types of fats and their chemistry

The main component of edible solid fats and oils is triglycerides formed by glycerol (a 3-carbon alcohol) combined with fatty acids.

---

[2] El Rico Colesterol, Andalucía Información, Libro de Corazón 2020

The lesser components of solid fats and oils are monoglycerides and diglycerides (one or two fatty acids with glycerol), free fatty acids, phospholipids (phosphatides), fat soluble vitamins, and cholesterol.

Triglycerides

The major component of edible fats, triglycerides, are formed by an alcohol with three carbons and three alcohol radicals $(OH)^-$, one in each carbon.

The simplified formula of the alcohol (glycol) is $C_3H_8O_3$ or $CH_2OH-CHOH-CH_2OH$ where the OH is the alcohol radical and is one negative electrical charge.

The glycol combines with a fatty acid, which is an organic acid with a COOH-R. When the alcohol (glycerol) combines with the fatty acid, the alcohol frees the OH^- radical and the acid frees the H^+ to form a molecule of water and a compound called ester (fat). When all three OH^- radicals have been replaced by fatty acids, the result will be a triglyceride (three esters) and three molecules of water.

Fatty acids

Fatty acids are organic acids formed by a chain of carbons that make one different from the next, and a COOH radical (organic acids group) characteristic of organic acid. The simplified formula for a fatty acid is R-COOH where R is the carbon chain (backbone of the fatty acids). This chain may be saturated where all carbons are linked by a single bond and have no double bonds (lard, butter, tallow). These are usually associated with high cholesterol content and are typical of animal fats, which are solid at room temperature. (and are usually associated with high cholesterol content in the ingredients from animals.)

Monounsaturated fatty acids

Have carbon chains with only one double bond C=C and when combined with glycerol form monounsaturated fats like olive oil, avocado oil, and macadamia oil. They are liquid at room temperature and the content of cholesterol in the olives, avocados, and macadamia nuts is minimal. This single unsaturated bond between the carbons allows for a configuration that facilitates enzymes to act in the molecule, as you can see in the diagram.

Molecule of oleic acid, predominant fatty acid in olive oil (Extra Virgin 76%)

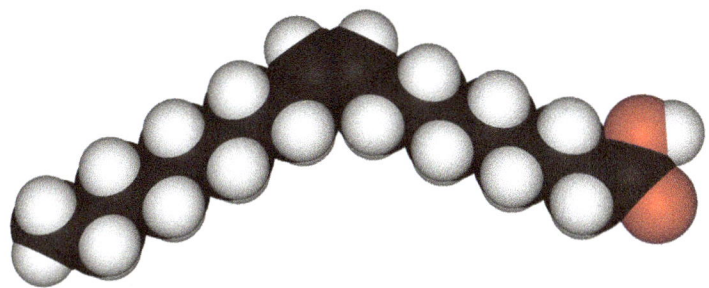

(Pub Chem oleic acid)

Note the aperture produced by the angle on the superior border, this facilitates the enzymatic activity in the molecule. Makes the monounsaturated oils the healthiest.

The following picture is the molecule of **stearic acid, a saturated fatty acid,** notice the straight alignment of the carbons. The lack of the angle like in the monounsaturated makes it difficult for the enzymes to metabolize and break down the carbon chain.

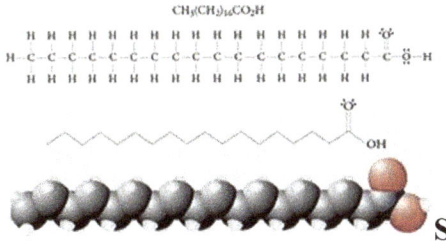 **Stearic acid, a saturated fatty acid.**

(Pub Chem stearic acid)

Polyunsaturated fatty acids

Have more than one double bond in the R-radical (carbon chain). Examples are all other vegetable oils like soya, corn, canola, peanut, etc.

The polyunsaturated fatty acids include Omega 3 and Omega 6 fatty acids.

Omega 3 fatty acids

Have anti-inflammatory properties and support all cell membranes and the production of hormones. They are alpha-linoleic acid (ALA), eicosapentaenoic acid (EPA) and docosahexaenoic acid (DHA) and are present primarily in fish, flax seeds, and fish oil as a supplement. Fish and their content of omega 3 in decreasing order are mackerel, salmon, trout, and sardines. Plant sources include flax seeds (ground), and oils from flax seed, canola, soybean, and nuts (walnuts).

Omega 6 fatty acids

Are pro-inflammatory, and when cooked will oxidize and promote the inflammatory process. They are found in vegetable oils, including soya and canola oils.

Our bodies can synthetize most fatty acids, except the essential fatty acids linoleic and linolenic acids that must come from dietary intake.

Trans fats

Are considered the worst types of fats because of their causative relationship with heart disease. They are **almost nonexistent in natural foods**. Minute amounts can be found in the milk of ruminants (cattle and sheep), and in breast milk of mothers who get the Trans fats from their diets. Most of the Trans fats from processed food products go through an industrial process called partial hydrogenation that makes vegetable oils into solids, such as shortening and margarine. It should also be noted that Trans fats can be produced by frying or heating vegetable oils (mono and polyunsaturated) at temperatures beyond their smoking temperatures or if oils are reused.[3] (Swati & Santosh, 2016)

Sterols

Are found in both animal and plant lipids, but with a big difference. The plant sterols called phytosterols (highest content being in soy and corn oils) have the property to lower the low density (bad) cholesterol. The sterols from land-based animals are primarily cholesterol. The content in the fats of beef, lamb, etc., is in the thousands of parts per million, while in fish is at about 100.

Phospholipids

Are a class of lipids indispensable for life, even though they are not of important nutritional value. They contain a hydrophilic head (water soluble) and a fatty acid tail that is hydrophobic (not soluble in water). This structure is ideal to form the cytoplasmic membranes of every cell in our bodies. They have an emulsifying property as well because of the head being water soluble and the tail being lipo-soluble, allowing it to hold mixtures of watery liquids and oils together, and acting as a bridge. Phospholipids are found in egg yolks and can be extracted from soya beans. This allows for preparations such as mayonnaise, Caesar salad dressings, and many other sauces, where water soluble ingredients (vinegar) are mixed with oils or lipo soluble ingredients.

Absorption of lipids

- **Triglycerides:** As we saw above, triglycerides represent the major component of the ingested lipids in our diet. They are not soluble in an aqueous medium, so they must be changed so they become soluble.

- **In emulsification**, large particles of triglycerides are broken physically and held in suspension. This happens through the effect

---

[3] Swati Bhardwaj, "Effect of heating/reheating of fats/oils, as used by Asian Indians, on *trans'* fatty acid formation," accessed June 23, 2021, https://doi.org/10.1016/j.foodchem.2016.06.021

of biliary acids present in the bile and secreted into the upper small intestine.

- **Triglyceride molecules** must be enzymatically digested into monoglycerides and free fatty acids. Both can be diffused and transported into the small bowel cell (enterocyte) through the effects of a pancreatic enzyme lipase.

Once inside the enterocyte (small bowel cell), the fatty acids and the monoglycerides are converted back to triglycerides and assemble with other fats forming particles called **chylomicrons** that are extruded from the cell into the lymphatic system that will transport them to the bloodstream after filtering them in the lymph nodes.

Absorption of cholesterol.

In addition to the cholesterol in our diets, there is a large amount of cholesterol that comes from bile. Cholesterol homeostasis is the result of a balance of ingested cholesterol (20 percent), synthesis in the liver (80 percent), and excretion of cholesterol in the bile. Cholesterol is readily absorbed through a specific transport protein (NPC1L1, Niemann-Pick C1-Like 1) that transports cholesterol from the intestinal lumen into the enterocyte.

Once in the enterocyte, the cholesterol is integrated in the **chylomicrons** with the triglycerides and others to be transported to the bloodstream through the lymphatic system.

Smoking points of some fats and oils.

Depending on the temperature at which you want to sauté or fry the food product you are preparing, you may choose an oil with a smoking point above the temperature required for your dish.: Smoking points of some fats and oils.

Wikipedia gives a very complete list of smoking points of fats and oils; we have selected the most commonly used ones.

Fat	Quality	Smoke point[caution 1]	
Almond oil		221 °C	430 °F[1]
Avocado oil	Refined	271 °C	520 °F[2][3]
Avocado oil	Unrefined	250 °C	482 °F[4]
Beef tallow		250 °C	480 °F
Butter		150 °C	302 °F[5]
Butter	Clarified	250 °C	482 °F[6]
Castor oil	Refined	200 °C[7]	392 °F
Coconut oil	Refined, dry	204 °C	400 °F[8]
Coconut oil	Unrefined, dry expeller pressed, virgin	177 °C	350 °F[8]
Corn oil		230–238 °C[9]	446–460 °F
Corn oil	Unrefined	178 °C[7]	352 °F

Fat	Quality	Smoke point[caution 1]	
Cottonseed oil	Refined, bleached, deodorized	220–230 °C[10]	428–446 °F
Flaxseed oil	Unrefined	107 °C	225 °F[3]
Grape seed oil		216 °C	421 °F
Lard		190 °C	374 °F[5]
Mustard oil		250 °C	480 °F[11]
Olive oil	Refined	199–243 °C	390–470 °F[12]
Olive oil	Virgin	210 °C	410 °F
Olive oil	Extra virgin, low acidity, high quality	207 °C	405 °F[3][13]
Olive oil	Extra virgin	190 °C	374 °F[13]
Palm oil	Fractionated	235 °C[14]	455 °F

Fat	Quality	Smoke point[caution 1]	
Peanut oil	Refined	232 °C[3]	450 °F
Peanut oil		227–229 °C[3][15]	441–445 °F
Peanut oil	Unrefined	160 °C[3]	320 °F
Pecan oil		243 °C[16]	470 °F
Safflower oil	Unrefined	107 °C	225 °F[3]
Safflower oil	Semirefined	160 °C	320 °F[3]
Safflower oil	Refined	266 °C	510 °F[3]
Sesame oil	Unrefined	177 °C	350 °F[3]
Sesame oil	Semirefined	232 °C	450 °F[3]
Soybean oil		234 °C[20]	453 °F
Sunflower oil		227 °C[3]	441 °F

Fat	Quality	Smoke point[caution 1]	
Sunflower oil	Unrefined, first cold-pressed, raw		225 °F

Chapter 4

Behind the Scenes – Understanding Cooking Methods

"Cooking is not effortless. To get a recipe that feels effortless is really hard."
—Claire Saffitz (Saffitz, n.d.)

Cooking is accomplished by the heat transfer from a source to food products. This transfer of energy occurs through air, water, and oils, and is due to collision of fast-moving or vibrating molecules or particles of these mediums with slower ones. Most cooking methods use all of them, but there is a predominant one used to segregate the cooking methods in the table below. Then we further segregate them by the form of heat transfer.

Types of heat Transfer

Conduction is by direct contact, usually between the heat source, a conductor (pan of some kind) and the food product.

Convection uses the phenomenon of energy absorption by a fluid (liquid or gas/vapor), which causes the heated portion of the fluid to rise and sink as it cools off. In the kitchen, it usually occurs in a closed space like an oven where the vapor will be the fluid in simmering or air will be the fluid in roasting dishes.

Radiation is the transfer of pure energy that warms up the receiving product, like the earth heated by the sunlight. Cloud cover will block the infrared portion of the spectrum (heat) but not the ultraviolet portion, so it's still possible to get a sunburn on a cloudy day while not feeling the heat.

For cooking, radiation has been used for a long time, a classic example being cooking over an open fire, placing the food product downwind from the fire so the radiated energy, and not the open flame, is used to cook. Modern ways of using radiation of heat include microwaves and infrared grills.

Each cooking method has a unique effect on food. We have talked about the effect of heat on each of the macronutrients of the ingredients we may use preparing our dishes. In general, the ingredients contain all the macronutrients, and they will change with heat; sugars will melt but not evaporate; starches will gelatinize; proteins will denature and agglutinate; fats will melt and if overheated they will smoke and change the nature of their fatty acids and form trans fats.

Additional changes occur when food products (ingredients) are subjected to special conditions created by a cooking method. They are the browning reactions of caramelization and the Maillard reactions, a series of reactions between the proteins and amino acids from the proteins breaking down and the sugars present in the ingredient being cooked. Meat, as an example, contains proteins and amino acids, carbohydrates in the form of glycogen in the muscle fibers, and fats in the marbling between the muscle fibers and the fat around the muscle itself.

The browning reaction produces flavors, aromas, and appearances that are characteristic of the cooking process and do not enhance or slightly modify the flavor or appearance of the ingredient, like seasonings herbs and spices do. This browning reaction is non-enzymatic, different from the enzymatic process of spoilage.

The information in the table below describes some easily recognized methods, but others in the table are more obscure. We will include a short explanation of the less common methods and expand more on the most used methods.

Table3 **Heat Transfer according to its nature**

NATURE OF HEAT TRANSFER	FORM OF HEAT TRANSFER	COOKING METHOD
Dry	Conduction	Dry roasting, hot salt frying, searing
Dry	Convection	Baking, modern roasting, smoking
Dry	Radiation	Grilling, traditional roasting, rotisserie, toasting
Wet	High Heat	Blanching, boiling, maceration, par boiling, shocking
Wet	Low heat	Coddling, poaching, simmering, creaming, slow cooking
Wet	Indirect heat	Bain Marie, sous vide, steaming, double steaming
Fat based	High heat	Blackening, browning, deep frying, reduction, shallow frying, stir frying, sautéing
Fat based	Low heat	Gentle frying, sweating
Mixed medium		Barbecuing, braising, flambé, fricassee, plank cooking
Device medium		Air frying, microwaving, pressure cooking, pressure frying, thermal cooking
Non heat		Curing, fermenting, pickling, souring

Cooking methods – general recommendations

One important concept is to simplify your methods. There is power in simplicity. Master the basics first, and your cooking will go much more smoothly.

Remember that perfection is the enemy of success, and that can be applied to many areas of life in addition to cooking. If you have a clear purpose for every step you take, you can avoid repeating steps because of errors on attempts at speed. Speed is not achieved by fast movements but by the avoidance of repeating steps because of errors. Unnecessary or incomplete steps complicate the process with no gain.

Divide the process into segments that are repetitive and easy to master. In surgery, the doctor organizes the work area where surgery is performed, stops the bleeding (hemostasis), and ties sutures. The cook hones his or her skills in using a knife, pairing ingredients, and cooking methods. Like the doctor, the cook also organizes the place of work and has all ingredients measured and prepared (**Mise en place**), a process like what a surgeon does in checking the instrument table in the operating room.

An early error does not necessarily ruin the day; correct it and move forward. Early fumbles clear the playing field for great finishes.

Now let us look at some cooking methods that will serve you well as you prepare your dishes.

Less common Cooking methods

Dry roasting

Is a process by which heat is applied to dry foodstuffs without the use of oil or water as a carrier. Unlike other dry heat methods, dry roasting is used with foods such as nuts and seeds. Dry-roasted foods are stirred as they are roasted to ensure even heating and can be roasted in a sauté pan or a wok.

Hot salt frying

Is done in a block of salt, heated in an oven, then the oil and food product are placed in it, and it goes back to the oven. I have seen it done and tried it myself, using an appropriate-size river stone brought to tableside to fry steaks on it. The problem with this technique is that you need to keep the stones hot all the time, as they will crack when cooling down. It works for restaurants, but for the occasional home cook who might do it on special occasions, the procedure is too onerous.

Traditional roasting

Is done on an open fire. It uses radiation from an open fire, still used today for the Asado Argentine style. A fire is built upwind from the cooking food product so the food (in an open environment) is not directly over the fire, but the wind will direct the heat to the cooking food. If there is a need for additional heat, the embers of the wood fire could be moved closer to or under the meat or other food product. This allows for better control of the cooking process, to obtain a Maillard reaction and avoid charring of the surface.

Maceration

Is the process where solids (meats, vegetables, fats) are cooked to flavor the cooking liquid. At the end, you may discard, or not, the solids and serve the liquid. This method is used to prepare broths, soups, etc. **Infusions** are a similar process for beverages (steeping) like tea, coffee, and chocolate.

Coddling

Coddling an egg is like poaching an egg but instead of cooking the egg in the water/vinegar mix, you place the egg in a recipient (ramekin) seasoned and then placed in the hot water to cook. Alternate methods include placing the ramekin in the microwave for 1- 1 ½ minutes at the lowest power possible, in my microwave is 1 minute at 40% power to have the white cooked and yolk a bit runny. Alternatively placing the egg in a plastic bag, eliminating all the air by submerging the plastic bag in water just before zip locking it, and then using the sous vide equipment (*Sous vide cooking.*

Double steaming,

Sometimes called *double boiling*, is a Chinese cooking technique[4] (Wikipedia.org, n.d.) to prepare delicate food such as bird's nest soup[5] and shark fin soup.[6] (Wikipedia.org/wiki/shark fin soup, n.d.) Cover the food with water and put it in a covered ceramic jar. The jar is then steamed for several hours in another pan with water. By doing this, the temp for the outer container is limited to 212°F or 100°C (boiling water temperature) and will be steady at that temperature as long there is water in the outside container. The temperature in the inner container will never exceed that temperature and will be somewhat lower because of the insulating effect of the ceramic container.

More common methods

Broiling

Is the current version of roasting over an open fire or hot coals (traditional roasting). Broiling uses radiation as the principal mode of heat transfer. All the sources of heat for broiling emit visible light so they are radiators of infrared radiation. The location of the source of heat is above the food product and the distance from the food to the source determines how deep the cooking process will go. This can be difficult, and practice will be necessary. The energy radiated from a broiler is 70-80 times as much as an oven in a baking mode. This allows a fast surface browning or caramelization (e.g., Crème Brule), while the center may remain cold. For this reason, for a home cook, broiling is a good finishing technique for meats but he or she may need a different technique for cooking the center, like roasting, sous vide, or grilling.

---

[4] Wikipedia, "Chinese cooking methods," accessed date June 23 2021, https://en.wikipedia.org/wiki/Chinese_cooking_techniques
[5] Wikipedia, "Edible Bird's Nest," accessed date June 23 2021, https://en.wikipedia.org/wiki/Edible_bird%27s_nest
[6] Wikipedia, "Shark fin soup," accessed date June 23 2021, https://en.wikipedia.org/wiki/Shark_fin_soup

Example recipe for broiling:

Carob Crème Brule

Ingredients

2 c. heavy cream

3 Tbsp. Carob syrup (specialty food stores or online)

5 eggs yolks

1 tsp. vanilla

1 c. sugar

INSTRUCTIONS

1. Preheat the oven 300°F.
2. Heat cream and carob syrup for 8-10 minutes but do not let it boil.
3. Beat ¼ cup sugar with the egg yolks and vanilla until the yolks pale and the mixture thickens.
4. Add egg yolk mixture to the cream and carob syrup in a thin and slow stream while mixing continuously.
5. Place the mix in ramekins positioned for a Bain Marie in a pan with an inch of water.
6. Bake for 35-40 minutes. The edges should be set and the center still soft and will quiver when shaken.
7. Let them cool to room temperature, take off the Bain Marie pan and place the ramekins in the refrigerator.
8. Just before serving, cover the surface with sugar and place in the broiler on high until golden at the surface, or use a flame thrower. The custard should still be cold.

Baking and roasting

Involve the food being surrounded by a hot box. This method depends on radiation from the hot walls of the oven and convection of the air contained in the box. The air alternatively heats and rises and cools and sinks since the source of the heat is at the bottom. Typically, temperatures obtainable with this technique are in the range of 300-500°F (150-260°C). In today's lexicon, we talk about baking when referring to cakes, pastries, etc. and roasting for savory dishes.

General guidelines for roasting:

- o Select tender cuts of meat, like meats from the ribs or loin areas, tender cut from the legs e.g., top round, some tender birds, or fish, e.g., whole chickens or whole fish.

- o When vegetables are cooked whole, their skins should be scored to permit the steam from the cooking interior to escape.

- o A layer of fat or poultry skin customarily is retained for basting naturally as it cooks. If you want to avoid any animal fat, use an encrusting as an alternative to skin (egg wash, flour and breadcrumbs or panko, or ground nuts), is added to prevent drying of the product.

- o If you are adding vegetables or mirepoix to your roast, make the cuts proportional to the time of roasting or size of the roast. If you are adding towards the end of the roasting, cut the vegetables small so they will be fully cooked and/or caramelized at serving time.

The recipe for beef roast can be found later in the book in Chapter 6.

Boiling, simmering, poaching, and Steaming

It is used as the mode of heat transfer convection of currents in hot water with a temperature of 212°F (100°C) for boiling, which may vary slightly with added ingredients to the water, such as salt, vinegar, etc. Therefore, these methods cannot achieve browning reactions of the food products. For the Maillard reaction to occur the surface temperature must reach 300°F.

Also, the temperature required for boiling is in equilibrium with the local atmospheric pressure of the location and as the altitude increases, the boiling temperature decreases, so that the boiling temperature in Denver is lower than in New York City. For each 1,000 feet of elevation, the boiling point of water drops 2°F. So, in Denver, with an altitude of a mile (5,280 feet) the boiling point will be 10.4° below 212 or 201.6°F.

For **simmering**, the temperature we aim for is 180-205°F (82-96°C) to avoid the violent bubbling of a boil to damage delicate ingredients.

 For **poaching**, we aim to reach 160-180°F (71-82°C) used often for fish. If the final product is the protein (chicken, fish, etc.) begin the poaching in hot liquid, or if the final product is the broth, begin the poaching in cold liquid, and it will infuse more flavor into the liquid (as in soups).

Example recipe **Poached eggs**

INGREDIENTS

2-4 eggs, room temperature
2 Tbsp. white vinegar
1 tsp. salt

INSTRUCTIONS

1. In a 12-inch skillet or pan, covered, put the water on to boil. I like using a saucepan for added depth of water.

2. In the meantime, place the shelled eggs in small ramekins.

3. As soon as the water breaks into a boil remove the lid, water should go into a simmer as it cools, when in a simmer, add the eggs at opposite ends of the skillet or pan so they don't touch each other.

4. Cover the skillet and turn the heat off.

5. Wait for 3-5minutes, (according to your taste for how well are done) the eggs should be done. Resist the urge to peek earlier.

6. Season with salt and pepper if desired.

Poached egg on toast

Braising and stewing

Are a combination of cooking methods, sautéing for the Maillard browning reactions, and then simmer. Therefore, temperatures are lower than the boiling point of water.

Braising

The protein is often cooked in serving-size portions, is sautéed, and then the vegetables and aromatics (your choice according to pairings to the protein) are added. At this point, add the liquid (water, stock, wine, etc.) to cover the protein to 1/3 to ½ of its thickness. Cover and cook low and slow, simmer until recommended inner temperature of the protein is reached.

Recipe example

Fricassee of grouper

INGREDIENTS

4 pieces of grouper filet about 8 oz. each
2 Tbsp. of olive oil
Salt and pepper, freshly ground
2 Tbsp. unsalted butter
Onion, carrot, and celery for Mirepoix (all medium size and cubed medium)
8 mushrooms, sliced (Shitake preferred)
2 Tbsp. all-purpose flour
½ c. white wine

2 c. fish stock, low sodium

1 c. heavy cream

2 Tbsp. chiffonade parsley

2 tsp. fresh thyme leaves

INSTRUCTIONS

1. Sauté the fish brushed with olive oil, seasoned with salt and pepper in a skillet with small amount of olive oil, just enough to cover the surface of the skillet, until golden; flip them over and repeat. Set it aside.

2. In the same pan add butter, melt it, and then prepare the mirepoix adding the mushrooms as well. Cook for 5-7 minutes and add flour (to create a roux); cook for a minute, stirring, then add the wine, scrape the bottom of the pan to loosen the fond and get the flavors of the sauté bits stuck to the bottom. Reduce the wine by half.

3. Add the fish broth, heavy cream, and season with parsley and thyme. Add the fish back to the pan and simmer to internal temperature of the fish of 122°F. Remove from heat and rest for 5 minutes.

4. Serve with rice or mashed potatoes to combine all the flavors.

Stewing

Involves ingredients that are usually cubed or cut into bite-size pieces, the proteins browned as in sauté, and then vegetables, aromatics, and the liquid of your choice are added to cover the food product entirely. Cover and cook low and slow to the recommended inner temperature of the food product. The difference between braising and stewing is the amount of liquid used.

Glazing

In cooking, is a coating of a glossy, often sweet, sometimes savory, substance applied to food. Often used in the preparation of pastries etc. Here will use it primarily for cooking vegetables in a mixture to cover the vegetable with a glaze to emphasize a flavor of the ingredient.

Example recipe Glazed carrots.

Ingredients:

- 1 lb. fresh carrots, peeled and cut evenly in ½" segments.
- 1 ½ Tsp. butter
- 1 1/2 Tsp. sugar
- Enough water to just cover the carrots (do not overcrowd the carrots, use batches).
- Distilled vinegar 1/8 tsp.
- Kosher salt to taste.

Procedure:

1. In a large pan add the carrots, sugar, water, butter, and vinegar. Turn heat to medium high and let the mixture reduce to a simmer, move the pan around to turn the carrot pieces frequently. When the glaze is almost done check for doneness of the carrots.

2. There should be no resistance to a paring knife tip. If not sufficiently cooked add a bit of water, lower the heat, and cook until done.

3. The aim of the movement of the pan and evaporation is to achieve an emulsion of the butter and water to create the glaze.

4. If caramelization occurs, reduce heat, add a bit of water and a couple of drops of vinegar.

5. Other glazes that can be used are Truffle infused Balsamic reduction, or bourbon barrel aged Maple syrup. Other vegetables to glaze include pearl onions, turnips, small beets, parsnips, and fennel.

Blanching

Is used to par cook an ingredient, which is cooked partially in a boiling liquid or fat. These ingredients are cooked in a second method as in sautéing, with a goal of having different items end up cooked simultaneously.

Frying

Uses oil as a conductor of heat and it imparts some flavor. Depending on the amount of oil, some convection will participate in the transfer of heat. Because most oils can be heated at high temperatures, near double the boiling point of water, browning reactions will occur, but instead you could char and burn the surface while the interior might not be cooked yet (the interior water content of the food prevents interior temps from getting beyond the boiling point of water). There are different types:

Sautéing

Uses a minimal amount of oil, only a thin layer. The food product is maintained undisturbed until the browning desired is completed. Then the food is turned, and the browning of the other surface is completed. Depending on the thickness of the food, the interior might be cooked or not. If it is thick, a combination cooking method may be necessary, such as significantly lowering the heat, finishing it in the oven, or adding a liquid (wine, broth, etc.) to complete the cooking with braising or stewing. **Stir frying** is a common technique in Asian cuisine, where the food has been cut in small pieces, is cooked in a relatively small amount oil at high temperatures and agitated while cooking, obtaining a uniform browning reaction and full thickness cooking because of the size of the food pieces. At times, the oil in the pan may be allowed to catch on fire, intensifying the browning.

Example recipe Stir Fry chicken with vegetables

Ingredients:

- 1 tablespoon + 1 teaspoon vegetable oil divided use
- 1 cup thinly sliced peeled carrots.
- 2 cups broccoli florets
- 1 lb. boneless skinless chicken breasts, cut into 1-inch pieces.
- 4 cloves garlic minced.
- 1/4 cup low sodium chicken broth or water
- 1/4 cup soy sauce
- 3 tablespoons honey
- 2 teaspoons cornstarch with a Tbsp. water for slurry.

- salt and pepper to taste

Procedure:

1. Stir fry the vegetables in 1 tsp. of oil at medium heat. Reserve
2. Heat up to high. Stir fry the chicken cut into 1" pieces in a Tbsp. of oil with salt and garlic. Cook 3-4 minutes until browned and fully cooked.
3. Add back the vegetables, mix the honey, chicken stock and soy sauce and add it to the pan.
4. Add the cornstarch slurry and bring to a boil and cook to get the sauce to thicken.

Serve immediately over rice.

Pan frying

Uses a larger amount of oil, enough to cover 1/3 or so of the food product. It is used in some cuisines, particularly when the food product is irregular, e.g., a whole fish, to get an even browning and uniform cooking, preventing sticking to the bottom. **Deep frying** employs enough oil to immerse the food completely. Because the oil temperature is high, it will create browning and since the size of the food is usually bite- size, it gets fully cooked by the time the browning is complete. This technique uses the convection mode of heat transfer, as in the boiling technique.

Grilling

Is the closest to traditional roasting on an open fire. You can cook directly over the flame or use indirect heat by building the fire to the side or back of the food product. It is one of the more intense heats produced by the different types of cooking methods, so cooking is fast and needs vigilance. Many prefer to cook on the hot embers of wood or charcoal and not over the flames. This requires starting your fire earlier at the side of the cooking surface and having access to be able to move the embers to the right location at the appropriate time. As with roasting, selection of the food product is essential. With meats, make sure the item is not so large, that it will not be cooked completely before the surface is charred, same for turkey and leg of lamb.

Shell crustaceans are best cooked on their shell, and fish chosen should be firm, like salmon, tuna, corvina etc.

Start wood fires outside of cooking area

Move coals to cooking area with V shaped grilling rod

Avoid flare-ups from dripping fat. If the grilling grate has V-shaped rods, most of the fat will run on the V grooves and be carried away from the fire, which works well. Make sure the grilling surface is clean before starting the fire, and steel brushes will do the trick, or use ½ an onion to clean the grate. Use tongs instead of forks to turn the food and use the internal temperature to determine doneness. Indirect signs are firmness of the product, looseness of the joints when cooking whole birds, and color of the juices on the surface, which should be clear or maybe just a touch of pinkness to your taste.

If cooking with wood instead of charcoal helps to have a separate area for starting the fire and develop the coals to use for grilling.

By Device medium

Microwave

Uses electromagnetic radiation as the source of heat. This kind of energy, electromagnetic radiation, is dependent on the number of cycles (full wave) per second (cycles/sec). There is a range called the spectrum from radio waves AM and FM (about 10^5 100,000 a 1 followed by 5 zeros) to x-rays and gamma rays at about 10^{20}. Microwaves like kitchen appliances are in the 10^{10} range and can put into vibration polar molecules, which is the reason microwave ovens heat food products from the inside (water - polar molecules) content of all food products). Ideal for food products that will dry the outer layers in the oven. A classic example is the Vidalia onion appetizer demonstrated earlier.

Infrared heat appliances

Use infrared waves with a frequency of about $10^{12.5}$ that are strong enough to produce vibration on non-polar molecules, like proteins, carbs, etc. used in ranges and electric grills.

Sous vide cooking. (Baldwin, 2013)

When cooking any product, we aim to make it more palatable and to kill enough pathogens (bacteria, parasites, etc.) to make the food safe, which is called pasteurization. Killing 100 percent of these microorganisms is called sterilization.

Raw meat products, according to USDA, should not be at a temperature higher than 41°F or 5°C for more than 4 hours. Sous vide cooking (under vacuum) refers to the process of sealing the food in a vacuum-sealed plastic bag and then cooking it in a water bath at a very precise and constant temperature necessary to bring the center internal temperature to desired doneness, for the time required. This, in effect, should achieve pasteurization.

INSTRUCTIONS

1. Attach the sous vide cooker to a deep pot, e.g., a pasta pot. Set the temperature and timer to the recommended values for your desired level of doneness. Once the temperature has reached the appropriate level, go to the next step.
2. Put your food in the sealable bag and clip it to the side of the container so the food is submerged and cook it for the recommended time.
3. Finish by browning the surfaces by searing, broiling, or grilling the food.

*SET UP FOR SOUS **VIDE** COOKING*

Table 4 Water Bath recommended temperatures for proteins.

Medium-rare beef: 130°F-139°F (54.4°C-59°C)

Medium beef: 140°F-145°F (60°C-63°C)

Traditional "slow-cooked" beef: 156°F-175°F (69°C-79°C)

Extra-juicy tender pork: 135°F-145°F (57°C-62°C)

Traditional tender pork: 145°F-155°F (62°C-68°C)

Traditional "slow cooked" pork: 156°F-175°F (69°C-79°C)

Extra-rare chicken breast: 136°F-139°F (58°C-59°C)

Traditional chicken breast: 140°F-150°F (60°C-66°C)

Mid-cut fish: 104°F (40°C)

Traditional fish: 122°F-132°F (50°C-55.5°C)

Table 5: Recommended times according to thickness: food products thawed and coming out of refrigerator at 41°F (5°C)

Thickness	Cooking time hours: minutes	Pasteurization time to 140°F 60.5°C
2.75-inch 70mm	6:25	3:50
2.36-inch 60mm	4:45	3.00
1.97-inch 50mm	3:15	2:20
1.57-inch 40mm	2:30	1:40
1.2-inch 30mm	1:15	1:10
0.8-inch 20mm	0.20	0.35
0.4-inch 10mm	0.8	0.25

Air frying

Is done in a countertop kitchen appliance that uses a heating element and a fan in a compact and small space to create a convective current primarily of air and maybe some ingredients you may have placed in it (oil). It is quite like a convection oven, but because of the smaller size of the box, it heats to the cooking temperature faster and cooking times are shorter. The heating element is electric, situated at the upper part of the fryer, surrounded by heat reflectors that concentrate the heat. A strong fan is located above the heating element and pushes the hot air down into a compact box that contains the food. The size of the box creates limitations as well. It works well for 2-4 servings, but you will have to work in batches for larger parties.

Is easy to clean and many of the accessory parts are dishwasher safe. It works best for frozen foods meant to taste deep fried, root vegetables, chicken, beef, pork, lamb, fish, mollusks, scallops, and other vegetables (asparagus, broccoli, Brussels sprouts, carrots, cauliflower, eggplant, mushrooms, peppers). **Air frying does not work well** with some fresh vegetables, green leafy vegetables, fresh cheese, or battered foods.

A website, bluejean chef,[7] (Meredith, 2016)has charts for different food products with times and temperatures, e.g., asparagus sliced 1-inch, 400°F 5 minutes.

If you have a traditional recipe that you want to cook in the air fryer, reduce the temperature by 10 percent and the time by 20 percent. Open the drawer of the air fryer as needed to check for doneness. Most air fryers will start back where you left it when opening the drawer. Since

Air Fried Brussel Sprouts

not all air fryers are calibrated the same, you may need to adjust the temperature or the time. If you find that you need to add oil, use a spray bottle. If you use skinless chicken or other skinless birds and they turn out too dry, encrust those using flour-egg-breadcrumbs, in that order. Remember to press the encrusting material with your hands into the surface of the skinless bird (have the breadcrumbs stick well to the food) so the strong air circulation of the air fryer won't blow the encrusting around.

---

[7] bluejean chef, "Air Fryer Cooking Charts," accessed date June 23 2021, https://bluejeanchef.com/cooking-school/air-fryer-cooking-charts/

COMBINATION COOKING METHODS

When more than one cooking method is utilized for an ingredient, it becomes a combination cooking method. It most often involves sautéing and simmering in a liquid or fat. The temperature of the liquid determines whether the method is poaching (160-180°F, 71-82°C), simmering (180-205°F, 82-96°C), or boiling (212°F, 100°C).

The amount of the liquid determines whether it is roasting (almost no liquid), braising (liquid to 1/3 to ½ of the thickness of the product being cooked), or stewing (product fully covered by the liquid). In poaching, braising, or stewing, the product is sautéed for browning purposes before the liquid is added. In blanching or par cooking (boiling a short time, product not fully cooked) the product will be cooked a second time with a different method, e.g., sautéed, roasted.

Roasting:

We refer here to the modern roasting method in an oven or on the stovetop. The main components of a roast are:

Star ingredient, which is the main item in a dish: May be a protein or vegetables. For meats, especially beef, this method is better suited for tougher meat cuts because long cooking at lower temperatures will make them tender at the end, e.g., roasts like top round, top sirloin, bottom round, and eye of round.

Seasoning and flavorings: Salt and pepper as seasonings to taste (e.g., 1 tsp. /lb. for salt, ¼ tsp. /lb. for pepper). Any herbs and spices you may prefer such as onions and garlic, fresh or dried. You may want to use a bit of olive oil to make a paste for a rub if the shape of the roast does not lend itself to hold the herbs and spices properly.

Oven temperature: If the proteins are of a shape and size that would allow sautéing, do that; if not, start with a high temperature oven (425-450°F, 218-230°C) for a short time, 15-20 minutes, then drop the temp. to 325-330°F (163-165°C) and roast until internal meat temperature is 130°F or 54.5°C.

Try your own proportions and choice of flavorings and seasonings and write your own recipe when you are happy with the result, so you can repeat it.

Example recipe:

Beef roast (stovetop)

Ingredients

Star ingredient (main ingredient): 4-5 lb. eye of the round

Condiments and flavorings as a rub: 6 cloves of garlic, pressed, 4 cloves split, 1½ tsp. kosher salt or to your taste, 1 tsp. black pepper, 1 tsp. cumin powder, 1 tsp. red wine vinegar.

Cooking liquid: 4 Tbsp. cooking oil (canola or avocado), ½ cup red wine vinegar, ½ cup water.

INSTRUCTIONS

1. Rub the meat with the mix of condiments and let it rest for ½ hour. Stab and insert in the stab wound the split garlic cloves, one in each.
2. In a deep saucepan, heat the oil and sear the eye of round, all sides, and then add the cooking liquid.
3. Reduce the temp to low and cover the saucepan with a lid.
4. Cook for 2 hours or so to an internal temperature of 130°F or 54°C, rotating the meat every ½ hour and checking whether you need to add any additional cooking liquid. Remember the internal temperature will rise the last 10°F quickly.
5. When the meat is done, remove the meat from the pan and rest it, covered with foil.
6. While meat rests, separate the fat (and discard) from the juices to prepare the gravy by adding beef demi glaze or beef stock and thickening with corn slurry to desired consistency. Corn slurry is prepared with cornstarch and cold water 1:1 by volume (e.g., 1 Tbsp. cornstarch to 1 Tbsp. water). This can be incorporated into a hot liquid without forming any lumps.

Recipe Beef Roast (in the oven)

INGREDIENTS AND INSTRUCTIONS

Star ingredient: 4 lb. round roast, or Picaña (sirloin cap)

Seasoning and flavorings: Salt and pepper to taste (1½ tsp. kosher salt, 1 tsp. black pepper, 1 Tbsp. rosemary leaves, 1 Tbsp. thyme leaves, 2 stalks celery cut in small pieces). Make the rub and smear on the meat. Place the celery at the bottom of the roasting pan.

Liquid: 1 Tbsp. olive oil mixed with rubbing ingredients.

Oven temperature: 450°F for 15-20 minutes (to brown the surface of the roast), then reduce to 325°F and cook to internal temperature of 130°F about 2 hours.

Rest for 30 minutes, tented with aluminum foil.

Roasted Picaña

Example Recipe

Roasted chicken

Ingredients and Instructions

Star ingredient: 1 whole roasting chicken, 5-6 lbs.

Seasoning and flavorings: 2.5 tsp. kosher salt, 1 tsp. black pepper, 1 bunch thyme + 20 sprigs (leaves), 6 cloves of garlic, crushed or 2 Tbsp. garlic crushed, 1 large red onion, sliced thick, 2 carrots, cubed, 1 bulb fennel, tops removed, cut in chunks.

In the cavity of the bird, place the bunch of thyme, all the garlic and lemon slices. At the bottom of the pan, place the onions, carrots, fennel, and celery. Spray with olive oil, salt, pepper, and the leaves of the 20 sprigs of thyme.

Place the chicken on top, roast at 450°F or 232°C for 25 minutes, then lower temperature to 400°F or 204°C for additional 45 minutes, and at or to internal temperature of 145-150°F (63°C) at the breast. Juices should be clear at a cut between leg and thigh. Rest for 20-30 minutes with tented aluminum foil before serving.

Roasting vegetables creates loss of water and therefore concentrates the flavors, makes you wonder why boil them.

Ingredients and Instructions

Star ingredients: Root vegetables like potatoes, sweet potatoes, turnips, parsnips, carrots. Crucifers like cauliflower, Brussel sprouts, cabbage. Others such as onions, bell peppers, garlic.

Liquid: olive oil, use liberally

Seasoning: kosher salt

Temperature ± 425°F (218°C); do not crowd the vegetables to avoid steaming them. Roast until tender when piercing with a fork and the ends are turning crispy. It is better if roasting is overdone rather than under cooked.

The harder the vegetables are when raw, the longer the roasting time needed. If you are roasting several vegetables at the same time, place the harder ones at the beginning and progress by placing the remaining vegetables according to diminishing hardness.

Root vegetables need 30-45 minutes, as do onions. Winter squash needs 20-60 minutes, depending on the size of the pieces. Cruciferous vegetables need 15-20 minutes like bok choy, broccoli, brussel sprots, cabbage cauliflower, radishes, turnips etc., soft vegetables 10-20 minutes, and thin vegetables like asparagus and scallions 10-15 minutes.

Example

Recipe Roasted zucchini and squash

Star Ingredient: 1 Green zucchini and 1 yellow squash

Vegetable oil: 14 Tbsp. canola oil

Salt: kosher and Malton (finishing sea salt scales)

Seasonings and flavorings: 1 peeled tomato (medium size) cubed medium, I Tbsp. and 1 tsp. champagne vinegar, 3 Tbsp. olive oil, 1 pinch cilantro.

Procedure:

1. Cut the zucchini and yellow squash in half-length wise, score the cut surface in a cross pattern, salt the surfaces and rest for 20 minutes.
2. Heat the vegetable oil in a large sauté pan, place the vegetables cut surface down, cook for about 5 minutes, add one or two garlic cloves and ají amarillo paste if desired.
3. Place immediately in a 450° oven and roast for 10 to 15 minutes, zucchini should be soft with no resistance to piercing.
4. Transfer the zucchini and yellow squash to a paper towel lined plate to remove excess oil.
5. Prepare the vinaigrette with cubed tomato, champagne vinegar, olive oil, minced shallot, and minced cilantro leaves.
6. Plate with zucchini and yellow squash and pour the vinaigrette over them.

Roasted zucchini and yellow squash with tomato and shallot salad

Example recipe

Roasted carrots

Star Ingredient Carrots

- 1-2 bunches of small young carrots (more tender)
- 1 shallot cubed medium.
- 1 tsp. garlic minced.
- Juice of a lime
- 2 Tbsp. of vegetable oil of your preference

Procedure

1. Preheat oven to 420°F or 450°F (215-230°C)
2. In a large sauté pan heat the vegetable oil in medium heat or medium low.
3. When the oil is shimmering add shallots and garlic. Cook for a minute or so till fragrant.
4. Add carrots and lime juice. Stir the carrots to make sure they are evenly coated.

5. Place sauté pan in the center of oven and roast for 15', carrots should show no resistance to a knife tip.
6. Serve them hot and may chop in thirds to serve as side dish of a beef dish.

Oven Roasted Carrots with garlic and shallots.

Braising is a combination method that benefits from dry and moist cooking methods.

Steps:

- Foods are usually browned first as in sauté.
- Then a base is prepared (soffritto, aderezo, etc.).
- Then a flavorful liquid is added to cover the protein by 1/3 to ½ of the thickness.
- Usually cooked covered (after adding the liquid) at a low temperature in a simmer (180-205°F, 82 96°C).

The components of a braised dish may include:

- A protein
- A base preparation (soffrito or aderezo)
- A flavorful liquid

- Condiments and flavorings (often added immediately after the liquid)
- Simmering for an appropriate time to cook the protein fully in the stove or oven.
- A thickening agent for the sauce
- Garnishing

Variations on braised dishes

Braised dishes utilize the cooking method of simmering, although often there will be some browning of the protein in use by sautéing. Therefor often braising dishes use combined methods. The main ingredient is often a protein that may be browned by sautéing and then set aside under tented aluminum foil, for a period. Usually in the same pan the aromatics will be cooked to form a soffritto, aderezo, Trinity and/or a Mirepoix.

Then the simmering for the braising begins by adding a simmering liquid. This liquid may be beer, wine, and/or broth, or a combination of them. The amount of the liquid should be enough to cover the lower 1/3 to lower ½ of the protein.

On addition there will be ingredients for seasoning and flavoring such as herbs and spices, and garnishing's. The cooking time is often critical, the best way to determine when is done is using an instant read internal thermometer, the temperature that is right will depend on the type of protein being cooked. At this stage you may cool it down, refrigerate overnight and remove the solid fat from the surface before reheating.

The residual liquid that remains will become the sauce, using a fat separator recover the juices, they will need to be thickened, and the most common way to achieve this is by using a slurry of corn starch or potato starch, alternatively may use the reduction technic. You prepare the slurry by mixing cold water or wine with corn starch or potato flour in a 1:1 ratio by weight. The slurry should be added in incremental amounts and allow to thicken a bit. When it does not thicken any more the sauce is completed.

Recipes Examples:

Beef Short Ribs

Ingredients and Procedure

- 3 lb. beef short ribs, sautéed to golden brown in two Tbsp. veg. oil and set aside.
- 1 onion, cubed small.
- 4 cloves garlic, pressed.
- Cook the onions and garlic at medium low heat when onions are translucent return the short ribs to the pot.
- Add the simmering liquid of your choice, and then the seasoning and flavorings (Salt and pepper; 2-4 sprigs of rosemary and thyme). Get to simmer on the stove top.
- Then place in a heated oven at 325°F (163°C). Cook for 2-2 ½ hours or until the meat pulls of the bone easily.
- Chopped cilantro leaves for garnish. Serve with rice or mushed potatoes.

Lamb Shanks

Ingredients and Procedure

- 4-6 Lamb Shanks sautéed 2 at time to golden brown, in 2 Tbsp. of canola oil. Then set aside covered with aluminum foil. Sprinkle with ½ cup of flour at this point.
- Aromatics: 1 onion cubed small, 1 Tbsp. garlic minced or pressed, 2 carrots slice ½ inch thick. Cook in medium low heat slowly. When the onions are translucent return the Lamb Shanks to the pot.
- Add the simmering liquid (1 ½ cups of red wine, 2 cups of beef stock, 14 oz. of tomato puree or sauce, 2 Tbsp. of tomato paste.
- Add immediately the seasoning and flavorings (Salt, black pepper, 2 bay leaves, 1 sprig of rosemary, 1 Tbsp. of chopped parsley leaves, 2 tsp. of beef demi glaze.
- Bring to a simmer on the stove top, then to a preheated oven at 350°F (175°C) for 2 ½ hours. Meat should fall apart from the bones easily. Adjust the temperature to a slow simmer.

- Garnish with 1 tbs. of chopped parsley leaves.

Pulled Pork Shoulder with Succotash

Ingredients for pulled pork: for 8-10 servings.

• Proteins	one pork shoulder bone in 6-7 lb.
	Bacon or pancetta 6 oz.
• Fat	Olive Oil 2 Tbsp.
• Aromatics	One Onion red Cubed medium.
	Garlic 1 head minced or pressed.
	Seeds toasted and ground, fennel, coriander, cumin 1 tsp each.
• Simmering liquids	Chicken stock low sodium 1 ½ cup.
	White wine, dry 1 ½ cup.
	Better than bouillon 1 tsp.
	Mushrooms shitake sliced 1 cup.

Ingredients for Succotash:

• Produce	Lima beans dry 10 oz. or canned 12 oz.
	Red Beans dry or canned 1 can 15-16 oz.
	Red Bell Pepper one diced.
	Onion red sliced thick, sauté lightly (with crunch remaining).
	Corn kernels 10-12 oz.
• Fat	Unsalted butter or extra virgin olive oil 2 Tbsp.
• Condiments and flavorings	Salt and Pepper to taste.
	Thyme leaves chopped 2 Tbsp.

Procedure for Pork Shoulder:

1. Pre heat oven to 300°F.
2. Sear pork in olive oil in a large oven proof deep skillet. Set Aside.
3. In the same pan with residual fat cook bacon or pancetta to crispy. Set it aside.
4. In the same pan with rendered fat prepare the base (sofrito) cooking the onion, as it begins to soften add the garlic, when fragrant (±1 minute) add the pork, bacon, seasonings, flavorings, and simmering liquids. Add the beef better than bouillon and mushrooms. Bring to a simmer in the stove top, then transfer to the oven and roast for 4 hrs. or until the meat falls off the bone.
5. Remove the pork and bacon, reduce the sauce to a thickness of your liking.
6. Shred or pull the pork with 2 forks. Serve the pulled pork with the sauce, sided by the succotash.

Procedure for the Succotash:

1. Cook the Lima beans and the red beans according to directions of the packaging. If canned drain and allow to air dry.
2. Sauté the onions, bell peppers and corn kernels, when the onions begin to soften add the Lima and Red beans, the seasoning and flavoring, reduce the heat to low, cover and simmer for 3-5 minutes or until all are heated through.

Pulled Pork Shoulder with Succotash

Sweated Chicken (Chicken Poeling)

In this dish a little fluid is added, you allow the ingredients to cook in their own juices.

Ingredients and Procedure:

- 2 chicken thighs sautéed in 1 Tbsp. veg. oil, seasoned with salt and pepper. When chicken is golden brown, set aside and keep warm.
- In same pan, add 2 Tbsp. veg. oil; on medium low heat, cook 1 Tbs. ají panca, 1 Tbs. ají amarillo, 1 onion sliced, and ½ Tbsp. garlic, pressed. May add other vegetables like mushrooms, broccoli, tomatoes etc. timing their addition according to their cooking time).
- When onions are translucent add 2 cups of the simmering liquid (chicken stock, white wine, or beer or a mix of your taste). Bring to a simmer for 10 minutes or so, add the chicken and simmer for 15 minutes, check the internal temperature, it should be 165°F.
- Serve with rice Pilaf style, maybe some sliced scallions and strips of plain omelet.

Rice pilaf Peruvian style.

My preference is to use optimum water level, pilaf style and combination of boil and steam. The method is as follow. It works for long, short grain rice and for basmati or jasmine rice.

1. In a tea kettle heat the water to boiling point.
2. In an adequate size saucepan place, I Tbsp. of vegetable oil per cup of rice to be used, oil should be of high smoking temperature like canola or avocado oils (to prevent formation of trans fats from excessive heat). Add salt and garlic, then add the amount of rice chosen and toast it for a few minutes.
3. Then add the proportional amount of hot water needed, the ratio is 1.5:1 of water to rice by volume.
4. With the saucepan uncovered continue to boil the mixture until you can see the rice surface, will appear like little volcanos erupting from the rice surface. At this point reduce the heat to extremely low, cover the saucepan with a lid, and steam for 20 minutes.
5. Fluff the rice with a fork and serve.

Chinese-Peruvian fusion Chicken

Ingredients

- **Protein:** 1 whole chicken ± 4 lb. divided into 8 pieces (breasts, wings, thighs, and drumsticks), browned with 1 Tbsp. veg oil.
- **Simmering liquid:** 1/3 cup soy sauce; ½ cup water; 1 Tbsp. Tomato puree; ¼ cup dry sherry
- **Seasonings:** ½ tsp. ají amarillo; 1 clove garlic, crushed; 1 green onion, sliced small
- **Thickening agent:** corn starch slurry (1:1 by weight in cold water)

Procedure:

1. Brown chicken, when golden brown set aside and keep warm.
2. Add simmering liquid,
3. Add seasonings and flavorings.
4. Add back the chicken.
5. Simmer for 45 minutes,
6. Remove chicken and thicken sauce by reduction or adding a corn starch slurry).
7. Serve with rice.

Braised Monk Fish

Ingredients

- Protein: 1½ lb. monk fish or grouper
- Aromatics:1 onion cubed small; 3 carrots in 1" chunks; 1 fennel bulb in chunks, 1 lb. potatoes in chunks
- Seasonings: salt, Pepper, ají amarillo paste or powder to taste.
- Simmering liquid: 1 cup white or red wine; 2½ cups chicken or fish stock
- Thickening agent: Corn starch slurry if desired (1:1 ratio by weight).

Procedure:

1. Brown the fish, set aside.
2. Prepare base prep. With all aromatics and flavorings, cook in low heat slowly until onions are translucent, and carrots, potatoes, and fennel partially soften, but still with some crunch left.
3. Add the liquids simmer for 10 minutes, by lowering the heat to simmer, when the vegetables are almost ready (±10 minutes) simmer for another 10 minutes. Add back the fish and simmer 15 minutes, to internal temperature 140°F, will reach 145°F while resting.
4. Serve with rice pilaf style, and the potatoes and other vegetables from the braise. If the braising liquid to thin, thicken it with the corn starch slurry.

Braised Leeks

Ingredients

- 4-6 leeks, cleaned and split lengthwise.
- Aromatics: Garlic ½ tsp. pressed. Flavorings 2 tsp. thyme leaves, 1 bay leaf, 1 tsp. sugar or stevia.
- Braising Liquid: White wine, 1 cup.
- Thickening agent: reduction by ½ through boiling.
- Garnishes: ¼ cup parsley leaves chopped.

Procedure:

1. Clean and prep the leeks, only use the white shanks.
2. Heat the butter, Add ½ tsp. minced garlic to foaming butter, just before the leeks are added.
3. Cook leeks in the butter and garlic. Cook the leeks in the butter, cut side down until the butter is foaming for 1-2 minutes; then turn the leeks sprinkle salt on cut side and cook for another 1-2 minutes cut side up,
4. When the leeks and garlic have cooked for 2 minutes on the second side, add the flavorings and the white wine. Simmer covered for 35-

40 minutes over low heat. When done, remove the leeks and reduce to ½ the simmer liquid by boiling reduction.

5. Serve as a side dish with any protein of your preference.

STEWS

Stewing is a combination method that takes advantage of dry and moist cooking methods. It is like braising, with the same elements, but the difference is that the food is completely covered by the simmering fluid.

The protein is often browned by sautéing and cut into bite-size chunks. The stew is prepared in a single pot and all ingredients are cooked together, with the timing of inclusion in the pot related to the time it takes to cook each.

GENERAL INSTRUCTIONS

1. Brown the meat in a pan, cover the meat with flour so it will form a roux with the oil used for browning. Do not overcrowd the pan; work in batches and set the browned meat aside.
2. Add vinegar or wine to deglaze the fond, scraping the pan to loosen any browned bits. Add the broth, the set aside meat, and bay leaves. Bring to a boil, then reduce heat to simmer.
3. With the pan covered, simmer for 1½ hours, skimming the broth foam every so often. Add the vegetables – carrots and onions, and simmer for another 10 minutes. Add potatoes and simmer for another 30 minutes. If the stew is dry, add broth or water; if it is soupy, cook uncovered for a bit more.

We will use the recipes to follow, as we did with the variations for braising, to provide ideas for different stews. Use your creativity and the pairing tables to achieve the desired results. Do not forget to write down your recipes so you can repeat them with the same great success.

Description of STEWS

Very similar to braising, the differences are primarily that the proteins are cut into bite size and the liquid covers the ingredients completely. The components and the process are similar.

Variations of stewing: a minor change of vegetable creates a whole new dish.

Classic Beef Stew

Ingredients

- Protein: 1 lb. of stewing meat in 1-inch cubes, dusted in flour and browned in 5 Tbsp. of canola oil and sautéing. Set apart and cover with tented foil.
- Aromatics: 1 medium onion diced in large cubes, 5 medium carrots diced in ¼ inch rounds, 2 large baking potatoes cubed large.
- Liquid for stewing: 2 Tbsp. wine vinegar; 1 cup red wine; 3½ cups beef broth.
- Seasonings and flavoring: ¼ tsp. black pepper; 2 tsp. kosher salt; 2 bay leaves.
- Thickening agent: ¼ cup all-purpose flour

Procedure

1. Combine the flour and pepper in a bowl, add the beef and toss. Cook the meat a few pieces at a time (to avoid steaming the meat instead of browning) turning the pieces until beef is browned on all sides, add more oil as needed between batches.
2. Remove the beef from the pot, add some of the stewing liquids and deglaze the pan. Add the beef, beef broth, the remainder of the wine and vinegar, and the bay leaves. Bring to a boil, then reduce to a slow simmer.
3. Cover and cook, skimming the broth from time to time, until the beef is tender, about 1 1/2 hours. Add the onions and carrots and simmer, covered, for 10 minutes. Add the potatoes and simmer until vegetables are tender, about 30 minutes more. Add broth or water if the stew is dry. Season with salt and pepper to taste.
4. To serve Ladle among 4 bowls.

Goulash

Ingredients for 8 servings.

- **Protein**: 2 lb. lean ground beef

- **Aromatics**: 2 large yellow onions cubed large; 3-4 cloves of garlic pressed.
- **Stewing liquid**: 3 cups water, 30 oz. tomato sauce, 30 oz. diced tomatoes, 3 tbsp. soy sauce.
- **Seasoning and flavorings**: salt and pepper to taste, basil dry 1 Tbsp., oregano dry 1 Tbsp., 3bay leaves.
- **Side dish**: 2 cups of pasta of your choice.

Procedure

1. In a large heavy pan cook the ground beef at medium heat, until starting to brown.
2. Then add the aromatics stirring some until the onions are translucent.
3. Add the tomato sauce and diced tomatoes and the seasoning and flavorings. Bring to a slow boil at medium heat. At this point cover and lower heat to low for a slow simmer for 20-25 minutes.
4. Stir the pasta and simmer on slow heat for 20-25 minutes. Check the pasta, should be al dente.
5. Remove the bay leaves and serve in soup type dishes.

Beef Bourguignonne

Ingredients

- **Protein**: 8 oz. applewood smoked bacon, 2½ lb. chuck beef, cubed in 1" pieces, 1 Tbsp. vegetable oil of high smoking point for browning
- **Aromatics**: 1 lb. carrots in 1" rounds, 2 yellow onions, sliced, 2 tsp. garlic, minced, 1 lb. frozen whole pearl onions, 1 lb. fresh mushrooms, stalk discarded, tops thickly sliced
- **Stewing Liquid**: ½ cup cognac, 1 bottle dry red wine, 2 cups beef broth, 1 Tbsp. tomato paste.
- **Seasonings and flavorings**: 1 tsp. fresh thyme leaves, salt 1 tbsp. and pepper 2 tsp.

Procedure

1. Preheat the oven to 250°F.
2. In a Dutch oven heat the canola oil or light olive oil, brown the bacon but not crispy, set aside. Brown the cubed beef in batches, all sides should be browned. Set it aside and keep warm.
3. In the same pan add the carrots and onions, salt, and pepper, toss them and cook until onions begin to brown. Add the garlic and cook for a minute.
4. Add the cognac, burn the alcohol, and return the bacon and meat to the pot with any juices accumulated. Add the wine and enough broth to cover the meat. Add the tomato paste, cover the pot and place in the oven for 1 ¼ hours until the meat and vegetables are very tender.
5. Sauté the mushrooms in another pan, add to the stew when out of the oven and simmer for 15 more minutes.
6. To serve, toast rustic bread pieces, rub a sliced clove of garlic, place in the plate and spoon the stew over it.

Marine Stew (of Fish and shellfish)

Ingredients

- **Protein**: 1 Tail of Monk Fish, 4 pieces of grouper 4 oz. each, 8 large shrimp, 8 medium size calamari, 5oo gm. Mussels.
- **Aromatics**: 1 large onion, 1 large leek, garlic 1-2 cloves, 3 tomatoes ripe, 1 small red pepper, 1 lb. frozen whole pearl onions.
- **Stewing liquid**: 1 jigger of cognac, 1 cup of fish broth or shrimp broth.
- **Seasoning and flavorings**: Parsley 1 bunch, saffron a small pinch, 3 Tbsp. almonds
- **A piece of hard rustic bread.**
- **Olive oil, and flour for the fish.**

Prep. Of the Ingredients:

- Remove the spine of the tail of the monk fish and peel the shrimp and heads if available. Keep the spine and shrimp shells to prepare a broth, with a couple of cups of water, boil it for 30 minutes after the boiling began. Let it cool and blend it to very fine, strain it with cheese cloth to be added to the mussel's broth.
- Cook in a pot the mussels with ½ inch of water, when the shells open remove the beards and shells, reserve the water, strain it with cheese cloth for the broth, and set aside the mussels.
- Dust the fish with flour and sauté them in cooking olive oil. Set aside Do the same with the 4 pieces of Monk fish tail.
- Clean the calamari, separate the tentacles and bodies, dry with paper towel and sauté them in olive oil and cook completely until tender. Set it aside.

Procedure

1. In a Dutch oven heat cooking olive oil and add the aromatics except the tomatoes. Cook at medium low heat, to slow simmer until the onions are translucent. Add the tomatoes as concasse and continue to cook for another 20 minutes.
2. In the meantime, toast the almonds peeled without added oil. Then add the almonds, saffron lightly toasted, to a mortar. Add the fried bread and with the pistil ground all. Set it aside.
3. When the soffrito is ready add the reserved fish to the pot, the shrimp, and the fried calamari. Cook for a couple of minutes and add the cognac. Let the alcohol evaporate and add the fish stock and part of the mussel broth. Add the almond grounded mix and sprinkle the parsley leaves.
4. Cook for another 5-7 minutes stirring occasionally to mix well the sauce. Serve over boiled potatoes, rice, or pasta.

More Variations of Stewing

Chicken Stew Peruvian Style Estofado de Pollo

Ingredients

- **Protein:** ½ oz. mushrooms dry or fresh, one 3 ¼ - 4 lb. chicken cut in 8 pieces, optional ½ cup of chicken livers and ½ cup of chicken gizzards.
- **Aromatics:** 1 onion cubed, 6 cloves garlic, diced, 2 Tbsp. Panca chili paste, 2 Tbsp. tomato paste, 3 tomatoes concasse, 4 oz. Carrots, sliced
- **Seasonings**: salt and pepper, 1 bay leaf, 1 cup packed cilantro leaves
- **Stewing liquids**: ½ cup red wine, 4 cups chicken broth
- **Other vegetables**: 3 oz. peas, 4 white potatoes halved, 1 ½ oz. of raisins, red kidney beans canned, rinsed, and dried.
- **Side dish**: 1 cup of rice long grain
- **Fats**: ¼ cup canola oil

Procedure

1. Slice mushrooms into bite size pieces
2. Heat the oil in a Dutch oven at medium low heat, season the chicken pieces with salt and pepper and brown in batches on all sides. When golden brown, remove the chicken and set aside.
3. Add some more oil in the same Dutch oven, cook the onions and garlic some, add the ají panca and tomato paste, add the beans and cook low and slow so the sauce thickens (10'). Then add the bay leaf, the tomatoes, and the chicken pieces, add some of the simmering liquids to obtain the consistency of your liking. In a separate saucepan cook the rice pilaf style, add 1 tsp. of ground turmeric to the boiling water for color. Season with salt and pepper. (The potatoes could be added to the simmering ingredients, my preference is to parboil them and complete cooking them in a roasting pan at 400°F for about 40 minutes).

4. Bring to a simmer, add raisins, peas, carrots. Cover and simmer for 15' until the potatoes are tender. If cooking the potatoes separately simmer until the internal temperature of the chicken reaches the safe temperature (160°F).

5. Add the cilantro leaves. Adjust seasonings and serve.

    Stew of Chicken with Bean Escabeche and rice. Stewing Sauce not added for picture clarity.

Chicken stewed with bean escabeche roasted potatoes and rice.

Pepian of corn

Ingredients

-   **Fats:** 4 ½ Tbsp. Vegetable oil
-   **Aromatics:** 1 cup full of white onion cubed small. 4-5 Tbsp. Ají amarillo paste, 4-5 Tbsp. Cilantro paste.
-   **Main Ingredient** 2 cups White corn kernels blended.
-   **Seasonings:** salt and pepper to taste, 1 tsp. brown sugar
-   **Liquid:** Chicken stock low sodium.

Procedure

1. Prepare the soffrito: heat the oil in a heavy saucepan at medium low heat. Add the onions, ají amarillo paste and cook until onion is translucent (≈8"). Add the cilantro paste (2-3 bunches of cilantro leaves and EVOO to form a paste in the blender).
2. Add the blended corn stirring frequently. Add a cup of chicken stock, and if it appears dry from evaporation add more.
3. Pepian is ready when the taste of raw corn is not present anymore, about ½ hour. Add seasonings including sugar.
4. Serve as a side dish, often with a protein-based dish.

Quinoa Stew with sea food mix

Ingredients

For quinoa stew

- 2 cups Quinoa white
- 3 cups of water
- 1 cup farmers' cheese
- 1 cup mixed shrimp, scallops, squid

For Soffrito (aderezo)

- 1 small onion, cubed small.
- 1 tsp. garlic minced.
- 2 potatoes cubed.
- 1 cube cilantro paste.
- Salt and pepper.
- 2 tsp. ají amarillo.

To complete stew

- 2 cups vegetable stock
- ¼ cup of heavy cream

Procedure

1. Boil quinoa in 3 cups of water, when beginning to boil lower heat to medium. Cook for 15 minutes. Quinoa will become translucent, and liquid will have been absorbed.
2. In a different pot prepare soffrito, heat 2 Tbsp. of vegetable oil, add the onions and cook low and slow until become translucent, add garlic, ají amarillo salt and pepper, and cilantro paste. When ready:
3. Add quinoa, potatoes, vegetable stock and cook in a slow simmer until potatoes are tender, about 15 minutes.
4. Add cream, farmers' cheese, and mixed seafood at the 10 minutes mark of the simmer of potatoes or if desired may add with the seafood a cup of tomato concasse.
5. Serve hot. May add at last minute some frozen peas or chopped Basil for added color.

Stewed quinoa with roasted potatoes and medley of seafood

Ajiaco of potatoes

Often this preparation is used as a side dish.

Ingredients:

- 1½ lb. potatoes par cooked and cubed medium to large.
- 1 lb. farmers' cheese
- 2 eggs hard boiled
- 1 red onion cubed 3/8" size
- 2 tsp. of pressed garlic
- 1-2 tsp. of ají amarillo paste.
- 1 ½ cups chicken broth
- ¼ cup evaporated milk or ½& ½
- Salt and pepper to taste
- 1 Tbsp. parsley leaves fresh.

Procedure:

1. Prepare the aderezo with cubed onion, crushed garlic, ají amarillo paste at medium low heat.
2. When the onions are translucent add the chicken broth and the cubed potatoes. Cook until potatoes show little resistance to a knife point.
3. Then add the milk or half and half, and the farmer cheese crumbled. Stir with a whisk until you have a homogenous mix. Season with salt and pepper to taste.
4. Plate with a protein of choice and top with slices of hard boil egg.
5. Alternatively, top with a fried egg.

Ajiaco of potatoes with spicy shrimp

As you can see, changing one or more of the elements of braised or stewed dish will produce an entirely different dish. *The methodology remains constant; therefore, you can repeat them or change them with confidence.*

Moist cooking methods are likely to produce good dishes because the food products are not easily damaged by high temperatures or too long cooking time.

Be confident and allow your creativity to guide you to discover new dishes.

Casseroles

Different cultures and ethnic groups have several baked one-pot dishes that could have been considered casseroles. The French popularized them and coined the name several centuries ago. Casseroles were adopted later in England and in the late part of the 1800s reached America, where they found a new home.

Today a casserole is any dish made of multiple ingredients that are baked together to create something all its own.

The term *casserole* is used for the cooking vessel in which it is prepared, is fire- and/or oven-proof, as well as for the dish prepared in it. Usually, it is served in the same cooking vessel it was baked in. Generally, casseroles that contain grains, pasta, or rice are baked covered at least part of the time. Casseroles made with cooked ingredients are baked uncovered, making the top crispier and browner. The fact that often the ingredients are pre-cooked makes casseroles an ideal way to use leftovers, especially if there are not enough to feed all, by using extenders.

The main parts of a casserole are:

- **The base** of the casserole provides its main texture and flavor, like cubed meats (beef, poultry, or fish), ground meats browned and drained, cheese, grated or in small cubes.

- **The extender** is a food ingredient that helps thicken a dish, like breadcrumbs, panko, cooked diced potatoes, rice, grains, cooked pureed Leguminosae (beans).

- **The binder** is a liquid that holds the other ingredients together, like milk products, broth, fruit juices, soup, eggs, thickened sauces, or mayonnaise. The advent of canned soups and creams added convenience to casseroles.

EXAMPLES OF CASSEROLES

Italian Chicken Casserole

Ingredients

- **Protein**: 1 cup cooked diced chicken, 6 cooked Italian sausages, ½ cup shredded
- **Dairy:** Parmesan Cheese, 2 cups mozzarella cheese, 2 Tbsp. Asiago cheese.
- **Extender**: 8 oz. penne pasta or other, al dente
- **Binder**: two 6 oz. spaghetti sauce
- **Vegetables**: ¼ cup onions cubed large, 1 cup diced Bell peppers, mixed

Procedure

1. Preheat the oven to 325°F.
2. Mix in a bowl all ingredients except mozzarella and asiago cheese.
3. Place the mix in an 11*7 inches baking dish, sprayed.
4. Top with Mozzarella and asiago, dust with the Parmesan cheese.
5. Bake for 40- 45 minutes until browned on the surface.

Beef Lasagna

Ingredients

- **Protein**: 1 ½ lb. ground beef 93% lean
- **Dairy**: ½ cup Parmesan, ½ cup mozzarella shredded
- **Extender / pasta**: 25 oz. frozen ravioli
- **Binder**: 26 oz. Marinara sauce, 1 cup of water
- **Vegetables**: 2 cups of mixed frozen vegetables, 8 oz. mushrooms sliced,
- **Seasonings**: ½ tsp. Kosher salt, ½ tsp freshly ground pepper divided, 1 tsp. ground cumin
- **Garnishes**: 1 cup French's crispy fried onions

Procedure

1. Heat the oven to400°F.
2. In a pan brown the ground beef with garlic ≈ 10', set aside and discard the fat.
3. In the same pan add the browned beef, remaining seasonings, vegetables, pasta sauce and water, cook for 2 minutes.
4. Coat a baking dish with cooking spray, make 2 layers of ravioli, then add the sauce – beef mixture, vegetables, and sprinkle heavily with cheese.
5. Bake covered for 20 minutes and uncovered for another 20 minutes.
6. Top with French's® crispy Fried onions and broil on high for 2 minutes
7. Rest and serve.

Beef Lasagna made with beef ravioli.

BEEF MUSHROOMS AND NOODLES CASSEROLE

Ingredients

- **Protein**: 1 lb. fully cooked beef tips with gravy
- **Seasoning / Flavorings:** 1 tsp. minced garlic, ½ tsp. dry thyme leaves, 1 Tbsp. soy sauce
- **Extender:** 3 cups egg noodles, cooked
- **Dairy:** 1/3 cup dairy sour cream,
- **Vegetables:** 2 cups frozen mixed vegetables, 8 oz. sliced mushrooms, French's crispy fried onions®.

Procedure:

1. Heat oven at 400°F, spray baking dish with vegetable oil spray
2. In a bowl, mix all ingredients, except the onions.
3. Place them in baking dish and put in the oven.
4. Bake for 35 minutes.
5. Top with French's crispy fried onions® and broil for 2-3 minutes.

Beef Shepherd's Pie

Ingredients

- **Protein**: 1 lb. ground beef
- **Seasoning / Flavorings**: 1 medium onion, cubed, ¼ tsp. black pepper freshly ground.
- **Extender**: 2½ - 3 cups of prepared mashed potatoes
- **Binder**: 12 oz. beef or mushroom gravy
- **Dairy**: 2 Tbsp. shredded Parmesan cheese
- **Vegetables**: 1½ cup frozen peas and carrots

Procedure:

1. Preheat oven to 450°F.
2. Brown ground beef and onions. Discard drippings ± 10 minutes
3. Place beef into baking dish, add seasonings, gravy, vegetables, top with potatoes and cheese.
4. Bake for 25-30 minutes until it begins to brown and bubble.

Now that we have talked about cooking methods, let us turn to some common preparations that will help you create dishes you will love to make and serve.

Risks of illnesses

All the cooking methods discussed in this chapter aside of making food tasty and pleasurable, must prevent contamination with pathogens (bacteria, contaminants) and the cooking methods used should not produce toxic substances like heterocyclic aromatic amines that are produced when excessive heat is applied to the food stuff to produce charring and smoky taste. The cooking method influences the production of these chemicals. Dry pan roasting produces a non-measurable amount, as do poaching, boiling, simmering, or stewing. Deep frying, grilling, and barbequing produce the most, when using these methods of cooking be careful not to allow the fats or proteins to get charred. (Oz & Kotan, 2016)

This consumption of heterocyclic aromatic amines is associated with increased risk of cancers of the pancreas, colon, lung, and prostate. (Payal & Ankita, 2018)

Chapter 5

Cooking Techniques and methods for the Secret Sauce

I've taught myself how to use good, fresh ingredients and to prepare them as simply as possible by cooking only to enhance their intrinsic flavors."
—Ina Garten (Garten, 2021)

In the present chapter, you will see information to help create delicious dishes using various ingredients and putting them together in new or old ways. I have included sections on using what I call "dish stars," a main ingredient that a meal can be centered on, and you will also find a few of these dish star recipes in Chapter 4 where they are used to illustrate cooking methods. Using variations in the process or ingredients changes your results. For the surgeon, the choice of an artificial valve for a specific patient is critically important. For the cook, using onions either raw, sautéed, or caramelized in a dish makes a big difference too.

But before we get to different ways to combine foods and flavors, let's talk about some cooking basic parts that will work no matter which ingredients you choose and how you combine them because they have been perfected over the centuries.

BASE PREPARATIONS

Many dishes begin with a base preparation and, depending on how you use them for broths, soups, or stews, you cut the ingredients in cubes, larger for soups or smaller for braises or stews. You cook them with small amounts of fat and the cooking should be done slowly at medium low amount of heat (low and slow). Basically, you sweat the vegetables on low heat and for a long time and then add a liquid to deglaze the pan (wine, broth, or water), in the amount and kind depending on the dish you want to prepare.

Mirepoix is made with fat, usually butter, and vegetables: 2 parts onions, 1-part carrots, and 1 part celery by weight and diced small to increase the surface and produce faster cooking. The vegetables should become soft and transparent.

Holy trinity is the Cajun version of the mirepoix but in it, the carrots are replaced by green peppers. The vegetable ratio can be equal parts of all three ingredients or 2 parts onions and 1 part of each of celery and green pepper. The fat is usually olive oil.

Pinçage is like mirepoix but it is cooked until it turns brown but not burned black, and then tomato paste is added. While mirepoix adds a clean, fresh vegetable complexity to your food, Pinçage is more assertive, has a strong presence, and is an excellent way to amp the flavor of stews and braises, or to use it on rice or quinoa salads, essentially enhancing umami.

Soffrito is a vegetable mix that varies in different regions, such as Puerto Rican soffrito that uses sweet red peppers, tomato concasse, onions, garlic, culantro or cilantro, and yellow pepper. In Peru it is called **aderezo** and uses onions, garlic, and ají amarillo.

Ratatouille sauce (≈ **caponata**) is a base preparation of creole origin and it uses in a 1:1 ratio yellow onion, green bell pepper, a small eggplant, and a medium zucchini, all diced. Add four minced garlic cloves, salt, and black pepper to taste, ½ tsp. of sugar, ¼ tsp. of Cajun spices, 15 oz. of tomato puree (15 oz.), 1 tsp. balsamic vinegar, and 3 Tbsp. water. In a large skillet, cook the onions and bell pepper low and slow in olive oil until the onions are translucent and bell peppers are soft, about 6-7 minutes. Add the eggplant, zucchini, garlic, salt, pepper, sugar, and Cajun spices. Reduce the heat to medium low and cook for 15 minutes, stirring occasionally. Add the tomatoes, vinegar, and water. Simmer covered for about 15 more minutes, stirring occasionally. Serve over pasta or rice or as a salad.

Caponata dish from Sicily, is remarkably similar. In addition to the Ratatouille sauce, olives and oregano are added and it is served as a salad.

Aderezo is the Peruvian version of a base, is prepared by cooking low and slow onions cubed small, garlic (pressed), and chili paste, (amarillo or red pepper –Rocoto-)

French mother sauces

These are called mother sauces because they are the origin of many variation sauces. They are prepared by the addition of a liquid to a thickening agent. (Delhindra, 2019)

Table 6: French Mother Sauces

NAME	LIQUID	THICKENING AGENT
Béchamel	milk	**white roux (butter + flour)**
Veloute	white stock	**white roux (butter + flour)**
Espagnole	brown stock	**brown roux + tomato paste**
Tomato sauce	tomato puree	**roux or reduction of tomato puree**
Hollandaise	clarified butter	**egg yolks**

The sauces should be thick; they should smother or cling to whatever they are drizzled, dolloped, or poured on. This consistency is accomplished by thickening agents like roux, by reduction like in tomato sauce, or by emulsification like in Hollandaise sauce.

How to make a roux: its two ingredients are fat (butter, ghee, or coconut oil) and all-purpose flour in a 1:1 ratio by weight. If you use 3 ounces of each this ratio will thicken a quart of liquid (milk, stock). **An ounce of all-purpose flour is about 3½ Tbsp., and an ounce of butter is 2 Tbsp**.

Melt the butter in a saucepan (remember, melted butter melts butter best) and add the flour progressively, stirring and cooking until you get a nutty aroma, indicating the flour is cooked (several minutes). The longer you cook it the darker it will get (yellow golden to brown and dark brown), but it will also lose its thickening power. Be careful not to burn it.

To prepare the sauce, add the liquid progressively (to avoid lumps), let it boil, and then turn off the heat when the desired thickness is achieved. You will be left with a thick, creamy warm mix.

Mother sauces

Béchamel sauce

Is a versatile sauce used in casseroles, as a base for savory soufflés, and with additions of other ingredients, becomes pasta or mac and cheese sauces, etc. (Saulnier, 1982)

Ingredients

6 cups milk or nondairy milk
2 oz. (4 Tbsp.) ghee or clarified butter.
2 oz. (7 Tbsp.) all-purpose flour

INSTRUCTIONS

1. Warm the milk, primarily to avoid splattering when adding to the roux.
2. Make the roux as above and cook to get the nutty aroma (blond roux).
3. Add the warm milk, a couple of Tbsp. first, stir with a whisk and when not lumpy, add the rest of the milk, whisk continuously, and cook until thick and creamy.

Veloute sauce

Includes chicken, veal, or fish Veloute sauce. It is not usually used as a finished sauce, even though it could be used as a gravy after seasoning it with salt and pepper. I like adding Peruvian ají amarillo (yellow pepper). Veloute sauce is most used as a base for many of its derivatives. The recipe that follows is used for the white wine sauce primarily used with fish and the base is a fish Veloute. (Saulnier, 1982)

INGREDIENTS AND INSTRUCTIONS

1. Warm the fish stock and keep warm (6 cups).
2. Prepare the roux as above using 4 Tbsp. clarified butter (2 oz.) and 7 Tbsp. flour (2 oz.). Cook until the nutty aroma emerges (blond roux).
3. Progressively add the warm fish stock, whisking vigorously until it is free of lumps. Simmer until it has reduced by 1/3, stirring so it will not burn on the bottom of the pan.
4. Remove the pan from the heat source and pour the sauce through a wire mesh lined with a piece of cheesecloth.
5. Cover the sauce surface with plastic wrap to avoid forming skin.

Hollandaise sauce (Saulnier, 1982)

Is most often used on Eggs Benedict, asparagus, salmon, and filet mignon, sometimes with added ingredients like chili sauce.

INGREDIENTS

4 Tbsp. butter
4 egg yolks
2 Tbsp. fresh lime juice
1 Tbsp. heavy cream
Salt and pepper to taste

INSTRUCTIONS

1. In a small saucepan or a double boiler, melt the butter, but do not let it brown.
2. In a bowl, beat the egg yolks, lime juice, heavy cream, salt, and pepper.
3. Temper the egg yolk mixture 1 teaspoon at the time with the warm butter up to 15 teaspoons (5 tablespoons).
4. Pour the egg yolk mixture into the melted butter, turn the heat to low, and whisk vigorously and cook for 10-15 seconds. If it is still

too runny, cook some more, a few seconds at the time until the consistency is to your liking.

5. Serve immediately.

Tomato sauce

Is used often with spaghetti, pizza, and other uses like shakshuka[8] (eggs cooked in tomato sauce)

INGREDIENTS AND INSTRUCTIONS

1. Big 4: garlic, olive oil, San Marzano tomatoes, basil
2. Mash up tomatoes, using a potato masher.
3. Salt liberally to counter the acid.
4. Use a cast iron skillet for preparing sauce.
5. Use sugar to counter bitterness, a pinch, or a tsp. of brown sugar.
6. Pasta: remove from boiling water 2 minutes shy of al dente. Finish it in the sauce mix and cover it.
7. Add a cup of the pasta water to the tomato sauce before adding the pasta.
8. Use dry herbs and spices (dry oregano, dry basil, and dry parsley). You may want to consider adding cumin, coriander, and/or ají).
9. Do not forget the dairy (cheese, for meats except chicken or fish).
10. Add wine and simmer to half the amount.
11. Add natural acids like lemon, capers, and olives if sauce is bland.
12. Balance the pasta with the sauce. The thicker the pasta, to balance add more acid.
13. Add more flavor to the sauce with EVOO at this point. Do not use EVOO for frying the meats; the high heat will turn the EVOO rancid, and the beneficial fatty acids will change to into Trans saturated fatty acids instead. Use light olive oil or other vegetable oil.

---

[8] Food52, "Eggs in Spicy Minted Tomato Sauce," accessed date June 23 2021, shakshuka[8]

14. Slow cook the sauce, preferably in a pressure cooker over low heat; simmer for 4 hours or so before adding the pasta. The sauce can be made a day or two ahead.
15. Just before serving add fresh chopped basil (use an extra sharp knife and don't press hard on the rolled-up leaves to avoid bruising the basil). To chop the basil, roll up the leaves, cut across to thickness desired and then cut them in opposite direction once.
16. Instead of grated Parmesan, consider shaved Parmesan on the side.

Espagnole sauce

Is a basic brown sauce that is the starting point for a demi-glace, a rich and deep flavored sauce used mostly with red meats. Is like Veloute sauce in that it is a stock-based sauce (beef stock in this case) and the thickening agent is roux (brown in this case). It has additional ingredients: tomato paste and mirepoix. (Saulnier, 1982)

INGREDIENTS

For Sachet:

Bay leaf
1/2 tsp. dried thyme
3 to 4 fresh parsley stems
7 to 8 whole black peppercorns

For roux:

1 oz. clarified butter.
1/2 cup onions (diced)
1/4 cup carrots (diced)
1/4 cup celery (diced)
7 Tbsp. all-purpose flour

Liquids:

3 cups brown stock (i.e., beef stock) unsalted
2 Tbsp. tomato purée

INSTRUCTIONS

1. Gather the ingredients.
2. Fold the bay leaf, thyme, parsley stems, and peppercorns in a square of cheesecloth and tie the corners with a piece of kitchen twine. Leave the string long enough that you can tie it to the handle of your pot to make it easier to retrieve it.
3. In a heavy-bottomed saucepan, melt the butter over medium heat until it becomes frothy.
4. Add the mirepoix—onions, carrots, and celery—and sauté for a few minutes until it is lightly browned. Do not let it burn, though.
5. With a wooden spoon, stir the flour into the mirepoix a little bit at a time until it is fully incorporated and forms a thick paste (this is your roux).
6. Lower the heat and cook the roux for another 5 minutes or so, until it just starts to take on a noticeably light brown color. Again, do not let it burn.
7. Using a wire whisk, slowly add the stock and tomato purée to the roux, whisking vigorously to make sure it is free of lumps.
8. Bring to a boil, lower heat, add the sachet and simmer[9] (Alfaro, 2019) for about 50 minutes or until the total volume has reduced by about one-third, stirring frequently to make sure the sauce does not scorch at the bottom of the pan.
9. Use a ladle to skim off any impurities that rise to the surface.
10. Remove the sauce from the heat and retrieve the sachet.
11. For an extra smooth consistency, carefully pour the sauce through a wire mesh strainer lined with a piece of cheesecloth.
12. If you will not be serving the sauce right away, keep it covered and warm until you are ready to use it.
13. Otherwise, serve hot and enjoy!

---

[9] The Spruce Eats, "How Simmering Is Used in Cooking," accessed date June 23 2021, https://www.thespruceeats.com/all-about-simmering-995786

Roasted beef with Espagnole sauce

You can use store-bought beef stock for making your Espagnole, but as always, make sure to use a low-sodium or, if possible, unsalted stock. Anytime you are reducing a liquid with salt in it, you will be concentrating the saltiness, which you might not want to do, especially if you plan to use the resulting sauce to make yet another sauce, which itself might be reduced. Better to season at the very end of cooking.

Derivatives of Mother Sauces *(Delhindra, 2019) (Saulnier, 1982)*

The mother sauces have a tremendous number of derivatives, too many for this simple cookbook, so will try to present them in as simplified way as I am able and at the same time provide enough information so you can create them to your taste, or maybe create your own new sauce.

Béchamel derivatives

Mornay Sauce: its origin is from the Béchamel, adding Parmesan and gruyere grated cheeses, there is no additional reduction needed, used in pastas and lasagnas.

Herb Sauce: its origin is from the Béchamel, adding herbs of your choice, there is no additional reduction needed, used in pastas, casseroles, and gratins.

Mustard sauce: its origin is from the Béchamel, adding prepared mustard, no additional reduction needed, is used on vegetables like cauliflower, broccoli, kale.

Soubise: its origin is from the Béchamel, added ingredient is sautéed diced onions or shallots, no additional reduction needed, used with vegetables, eggs, chicken, and casseroles.

Lobster sauce: its origin is from Béchamel, adding anchovy paste, diced lobster meat, ají. No additional reduction needed, used with all types of seafood.

Cream Sauce: Its origin is from Béchamel, adding heavy cream, no additional reduction is needed, used in sandwiches like Croque Monsieur and others.

Veloute Derivatives

Supreme sauce: its origin is from Veloute Chicken, adding cream, no additional reduction needed, used with poultry specially poached.

White wine sauce: its origin is Veloute fish. Additional ingredients include white wine with additional reduction at this point, adding then cream, butter, salt, and pepper. Used primarily with seafood.

Normandy Sauce: its origin is Veloute fish, additional ingredients (in equal volume as the velouté fish) of mushroom liquor, if using dried mushrooms and mussels' liquor or fish stock. Reduce by half, add the cream and (4 Tbsp.) butter by pieces until the desired consistency achieved. May use a combination of butter and egg yolks for liaison.

Sauce Bercy: its origin is the Veloute fish, with additional ingredients of white wine, shallots minced with additional reduction at this point to ½, then add butter 4 Tbsp. by pieces until desired consistency, flavor with salt, pepper, and other spices of your choice. Is used primarily with poached fish.

Mushroom sauce: for fish its origin is Veloute fish with additional mushrooms, reduce by ½ add a liaison of egg yolks and heavy cream to desired consistency. Is used primarily for fish and crustaceans.

Sauce Allemande or Parisienne: its origin is from Veloute veal, the additional ingredients include cream, egg yolks lemon juice salt and pepper. No additional reduction is necessary. Is used for eggs, poached chicken, veal, and vegetables.

Mushroom sauce for poultry: its origin is the Allemande sauce with additional mushrooms no additional reduction is needed. Used for poultry.

Bonnefoy or white bordelaise: its origin is the Veloute sauce fish or chicken, add Bay leaf, pepper thyme, reduce to half, add the white Bordeaux wine and shallots, reduce by half again. Is used for chicken, fish, shrimp, and lobster.

Sauce Indene: its origin is Veloute chicken, fish, veal. Additional ingredients include curry powder, coconut cream or milk, lemon juice or alternatively may use curry pastes. Reduce the mix to desired consistency. Can be used with any protein depending on the velouté used, can add potatoes or other vegetables as well.

Espagnole Derivatives

Sauce Bordelaise: its origin is Espagnole sauce, with added red wine, shallots, and beef demi glaze while reduction in process add bay leaf, thyme, and pepper. Begin by slow cooking the shallots in butter, when caramelized add the wine, lower the temperature let it simmer until wine is almost evaporated, add the Espagnole sauce or veal stock and cook until reduced by ½. Pour sauce through a conical fine mesh strainer to the very last few drops. Season with salt and black pepper.

Sauce Bourguignonne: its origin is in the Espagnole sauce and begin the process by slow cooking the shallots in clarified butter, add the red wine and cook to almost evaporated then add the Espagnole sauce and some chilies if you prefer. Cook to the preferred consistency.

Sauce Diable: origin in the Espagnole sauce. Needs additional reduction of added white wine and shallots to the Espagnole sauce. Used in grilled chicken or pigeons.

Sauce lyonnaise: origin in the Espagnole sauce. Begins with browning the onions and needs additional reduction to the added white wine, vinegar and the Espagnole sauce. Used in roasted meats, pork, poultry, and sausages.

Madeira sauce: has its origin in the Espagnole sauce. Needs additional reduction of the added Madeira to desired consistency. Used in roasted meats and steaks.

Perigueux sauce: has its origins in the Espagnole sauce with added ingredients of truffles and truffles pâté. Needs no additional reduction. Used in roasted meats and steaks.

Sauce Bercy (Brown): has its origin in the Espagnole sauce. Begins cooking the shallots in butter and crushed peppercorns. Needs additional reduction for the added white wine to the Espagnole sauce. Is used for grilled meats.

Tomato sauce Derivatives

Portuguese sauce: its origin is in tomato sauce. Needs additional reduction of the tomato sauce with added Tomato concasse, meat glaze, parsley, olives, anchovies, chopped onions, garlic, salt and sugar. Is used in fish sautéed fish and then simmered in the sauce.

Milanese Sauce: its origin is in the tomato sauce. Needs further reduction for added beef demi glaze. Begins with onions cubed small, garlic pressed, and juliennes of mushrooms until soft. Then add all the other ingredients and cook low and slow. Used in pasta, chicken and pounded and breaded beef.

Hollandaise sauce derivatives

Sauce mousseline or Chantilly: its origins is in Hollandaise sauce where you mix 2/3 of Hollandaise with 1/3 of whipped heavy cream. Is used in steamed or poached fish, asparagus, or broccoli.

Sauce Bavarois: its origins are in the Hollandaise sauce. Needs an Additional reduction for the added crayfish butter and crayfish tails. Is used on fish.

Sauce Rubens: has its origins in the hollandaise sauce. Will need an additional reduction. Begin with a reduction of white wine, fish stock and fine mirepoix, strain add yolk of eggs and crayfish butter to finish.

Béarnaise sauce: its origin is with Hollandaise sauce. Begins with a reduction of chopped shallots, mignonette pepper corns, tarragon, salt, and vinegar. Then add the egg yolks and melted butter whisking briskly on low fire to ensure the yolks get cooked, strain through a cheese cloth covered strainer. Finally add chopped tarragon and chervil. Is like hollandaise but the acid used is vinegar instead of lemon Juice.

Francoise sauce: its origin is the béarnaise with added fish glaze and tomato puree. Is used on fish and pasta.

On addition to these many sauces there are numerous sauces by ethnic and regional groups as well as sauces by their time-period. Examples are in a book titled Modern Sauces (Holmberg, 2012) where you will find as an example a recipe for "Caramelized Onions Coulis" which is prepared with caramelized onions, a reduction of crème fraiche a broth and some acid from lemon juice.

Cooking rice

Rice is cooked when it's temperature reaches a point when the crystalline structures of the starch begin to melt. This is called gelatinization temperature (GT) and it ranges between 131°F-55°C and 185°F-85°C.

Methods: Most of the methods depend on the proportion of water and rice. In general, the longer the rice grain, the fluffier the final product; the shorter the grain, the stickier the final product. Some methods use excess water, discarding the excess water after the rice is cooked; others use an optimal amount of water which varies according to the type of rice used. The rice may be boiled, steamed, or cooked using a combination of the two.

My preference is to use an optimum amount of water (1 cup rice to 1½ cups water on long grain), pilaf style and combination of boil-steam. The definition of rice pilaf is difficult because it is prepared in many ways around the world. We are using the term as basic rice pilaf made by sautéing, usually a long grain rice in butter or oil or both and some pressed garlic. Sauté till the rice browns slightly and then add whatever kind of broth you want or water. It works as well for short grain rice, basmati, or for Jasmine rice.

Example recipe Rice Pilaf method

Rice pilaf Peruvian style

INSTRUCTIONS

In a tea kettle, heat the water to boiling.

1. In an adequate-size saucepan, place 1 Tbsp. of oil per cup of rice to be used and a tsp. of pressed garlic per cup of rice. Oil should be of high smoking temperature like canola or avocado oils (to prevent formation of trans-fats from excessive heat). Add the amount of rice chosen and toast it for a few minutes. Add salt to your taste, e.g., 1 tsp.
2. Then add the proportional amount of hot water needed, the ratio is 1.5 to 1 of water to rice by volume.
3. With the saucepan uncovered, continue to boil the mixture until you can see what looks like little volcanoes erupting from the rice surface, or like molten lava. At this point turn the heat to extremely low, cover the saucepan with a lid, leave a little vent, and steam for 18 minutes.
4. Fluff the rice with a fork, turn the heat off, and serve.
5. If you want to add color to the rice may add ¼ tsp. of pimentón (smoked paprika) or of turmeric.

Time to cover and reduce heat.

Fully Cooked ready to be fluffed with a fork.

Example recipe

Chicken and rice, Peruvian style

Ingredients:

Herbs and spices: 1 cup cilantro leaves, or 1 Tbsp. cilantro paste (see recipe page 17).
1 tsp. ground cumin
1 bay leaf
Salt and pepper to taste

Aromatics: 1 small onion, diced small
3 cloves of garlic, minced.
2 tsp. ají Amarillo

Simmering liquid: 1 cup beer or white wine
1 cup chicken broth

Vegetables: 2 small carrots, diced small
1 small red pepper, diced small.

2 cups peas

1 cup corn kernels

5 scallions, chopped.

Rice, long grain   1 cup (basmati, jasmine is okay)

Chicken   4 chicken legs and thighs

Procedure

1. Pulse cilantro leaves and 2 Tbsp. of water till smooth.
2. Brown the chicken in pan with a bit of oil. Set aside.
3. Cook low and slow the onions, garlic, cumin, ají amarillo until the onion is translucent; do not brown.
4. Add the simmering liquids, scraping to free the fond, which is the residue at the bottom of the pan or skillet from browning the chicken.
5. Add the bay leaf.
6. . Add the sautéed chicken, vegetables, pureed cilantro, salt, and pepper. Simmer for 15 minutes, covered.
7. Add the rice, cover, and simmer for 20 minutes until the rice is done.
8. Add the peas and corn kernels if frozen for a couple of minutes before finished; otherwise add them with the rice.

Chicken and rice Peruvian Style

NOTE: You may replace the chicken for duck and have another delicious Peruvian dish Arroz con pato (rice with duck)

Rice with seafood

Ingredients

- **Herbs and spices**: ¼ cup cilantro leaves or one cube of cilantro paste.
- 1 tsp. dried oregano
- Salt and pepper to taste
- 1 lemon, sliced.
- ¼ tsp. pimenton (smoked paprika)

Aromatics: 1 Small onion, diced small minced, 2 tsp. ají Amarillo
Simmering liquid: ½ cup of white wine, 1 cup chicken broth.
Vegetables: Plump tomatoes concasse 2,
Protein: 1 lb. mixed (of your preference) seafood, raw
Rice: 1.5 cups long grain rice and 2¼ cups fish or shrimp broth
Fat: Vegetable oil with high smoking point (canola, avocado, etc.)

INSTRUCTIONS

1. Sauté the seafood mix and set aside.
2. Prepare the soffrito: 3 Tbsp. vegetable oil, heated, add the aromatics one at a time in the order they are in the aromatics list. Cook until the onions are translucent. Add the oregano, the cilantro paste, and a pinch of salt and pepper.
3. Add the wine, broth, and rice. Boil for 3 minutes, reduce the temp to low, cover, and simmer for 18 minutes.
4. At the 10-minute mark, add the seafood mix that was set aside and cover again. At the 18-minute mark, check for the seafood mix being done and adjust seasoning. When done, sprinkle with cilantro leaves and serve with a slice of lemon.

Paella method

Most often, one uses one or maybe two different sizes of sauté pans or paella pans to prepare paella. To get consistent results, determine the optimal amount of ingredients for each of the pans you frequently use.

The ingredients to be calculated include amount of rice, amount of stock or water, and how to make the stock.

1. Calculate the rice amount, from a thin layer of rice that covers the bottom of the paella pan. Weigh the rice or pour it in a measuring cup, and now you know the amount of rice for your paella pan.
2. Calculate the amount of stock or water necessary for the rice in your paella pan. Put 2 quarts of water in your empty pan, boil for 5 minutes, simmer for 10 minutes, and then boil for another 5 minutes. Then measure how much water is left. Subtract that from 2 quarts so you know how much evaporated. The stock or water necessary will be equal to three times the volume of rice, plus the volume of water evaporated.

3. Stock preparation: 2 dozen whole raw shrimp with heads on – shell the shrimp tail and put the meat aside. Sauté some garlic to taste, gently for a couple of minutes; do not let it brown or it will become bitter. Then add the shrimp heads and the shells; cook for 3 minutes and add good quality fish stock in the amount necessary for your paella (as measured for your paella pan) plus ¼ cup. Simmer for 30 minutes, then strain and measure. If it is short of what you need, add enough fish stock for the amount needed.

5. For other ingredients, use the reserved shrimp meat and add other seafood ingredients like squid, monk fish, etc. If desired, add chorizo, chicken, or whatever else you may prefer.

23 Thin Layer of rice in 16' sauté pan   *24 Residual water after boiling 1250 cc*

The rice in the pictures weighed 118g or 4.1 oz., It will absorb 3 times or 354g. or cc. of water The total evaporated water is 2000 cc- 1250cc equals 650cc.

Total liquid needed for that paella pan will be 354+650 equals 1 liter (≈1 quart), it may be fish stock, water and wine or a combination of those.

For the best results use heavy bottom pans or use double circular burners to achieve uniform heating of the classic paella pan.

Best paella recipe:

inspired by Cloake's perfect paella[10] (Cloake, 2011)

Ingredients

Proteins: 1 doz. whole raw shrimp, 5-1/3 oz. squid, 5-1/3 oz. monk fish, 5-1/3 oz. mussels

Aromatics: 1 onion, diced small, 7 oz. chopped tomatoes, 1 tsp. minced garlic

Condiments: Salt and pepper to taste, 1 tsp. pimentón (smoked paprika), 1 pinch of saffron in 1 Tbsp. hot water

Simmering liquids: 6 Tbsp. EVOO, 4 cups fish stock or shrimp stock prepared as above, 1½-2 fluid oz. white wine.

Rice: Short grain Bomba Calasparra as calculated for your paella pan, just enough to cover the bottom of the pan in a single layer.

Broad beans: 5 1/3 oz.

Procedure

1. Prepare the stock as in Paella method.
2. Heat the oil in the paella pan; sauté the monk fish to golden, but do not fully cook. Remove and set aside.
3. Add the aromatics, onion, and garlic first, cook until soft, add pimentón mix and cook for a minute. Add the tomatoes and wine and simmer for 10 minutes. Add the squid and beans.
4. Stir in the rice, mix well, and add 3½ cups of the stock, the saffron, and soaking water. Simmer vigorously for 10 minutes.
5. Arrange the monk fish, mussels, and squid on top of the rice and push them gently into the rice. Cook for about 8 minutes; if it looks dry, add the remaining stocks a bit at the time, but do not make it look soupy.

---

[10] The Guardian, Felicity Cloake, "How to cook the perfect paella," The Guardian, accessed date June 23 2021, https://www.theguardian.com/lifeandstyle/2011/aug/18/how-to-cook-perfect-paella

6. Rest it off the heat, covered in foil for 10 minutes. Then garnish with flat leaf parsley and squirt half a lemon over it.

Even though classic paella has rigid and strict recipes, and the purists insist that a specific type of paella must be done just so, it is still a dish that allows creativity and freedom to use what you have at hand. It is a wonderful and entertaining dish, allowing the freedom to pair some unlikely ingredients.

The next recipe is an example of things you can do with your creativity.

Seafood Paella, Larger pan

Paella with artichokes, chicken, chorizo, and shrimp

Ingredients

Sea food paella

- **Proteins**: 6 bone- and skin-in chicken thighs, 6 fresh chorizo sausages in 2" pieces, 12 shrimp (21-25 count/lb.).
- **Vegetables**: 2 fresh artichokes trimmed, steamed, and sliced lengthwise in quarters, fur removed, and 6 tomatoes concasse or canned Marzano type.
- **Aromatics**: 1½ onions diced small, 6 garlic cloves pressed, 1½ red bell peppers, diced small (uniform cut = uniformly cooked)
- **Condiments**: ½ tsp. pimenton de la Vera, 1 pinch kosher salt, ½ pinch black pepper, 1 pinch saffron in hot water
- **Rice**: Bomba/Calasparra, appropriate amount for your paella pan (see paella method above).

128

- **Broth**: Since you have 3 different types of proteins you can use shrimp broth or chicken broth or pork or even beef broth, whatever you have at hand; the amount appropriate for your paella pan (see paella method) + ½ cup
- **Garnishes**: flat leaf parsley leaves chopped, and the slices of 2 lemons

Instructions

1. Using your paella pan, sauté the chicken (do not overcrowd), then the chorizo, and then the artichokes, setting each aside when done.

2. Prepare the soffrito in the same pan, using the residual fat and adding olive oil as needed. Add the aromatics sequentially – onions, garlic, and red pepper, and cook until onions are translucent. Add the tomatoes and cook until the consistency is thick and syrupy. At this point add the condiments.

3. Assemble the paella:

   - Heat oven to 400°F.
   - Add the rice and stir (only one time) cook for a few minutes, making sure rice is spread in a thin layer at the bottom of the pan.
   - Add the hot broth as calculated for your paella pan.
   - Add and spread the chicken, chorizo, and artichokes on top of the rice; cook at medium low heat for 10 minutes after an initial short boil. Finally, add and spread the shrimp. If the rice looks dry, add, as necessary, the remainder of the extra ½ cup of broth.
   - Place it in the oven and cook for an additional 10 minutes. If too soupy, cook a few minutes longer.
   - Add slices of lemon.

Chapter 6

Creating a Tasty and Flavorful Dish

"I cook with wine, sometimes I even add it to the food."
—W. C. Fields (Fields, n.d.)

To create a dish, it is necessary to mix two or more ingredients to create a pleasant sensation to the people dining. Not all ingredient combinations accomplish this, but the right ingredient combined with other right ingredients in the right proportions may produce a dish where the result is greater than the parts.

This may sound difficult, and it may be, but with the proper tools it can be simplified.

Remember that in the sense of taste one will enhance or balance others; we need to keep this in mind when selecting our ingredients. Do we want to enhance or balance their tastes, or do we want to use additive power by using ingredients with similar taste, or do we want to create contrast by adding an ingredient quite dissimilar?

By using varying flavors and ingredients, you can create a flavor identifiable with the dish, giving it almost a personality all its own.

With millions of recipes worldwide and the power of software and computers, multiple websites have produced lists of pairings. Such examples can be found on the Food Pairing[11] (https://inspire.foodpairing.com/account/login, n.d.)

Most often a home cook on a mission to produce dinner in the shortest time after a long day at work needs a quick way to decide what to cook and complete the process. Maybe opening the refrigerator to see what is there, the cook will choose an ingredient that will be the star of the dish, often a protein

---

[11] Food Pairing website, accessed date June 23 2021
https://inspire.foodpairing.com/Account/Login?ReturnUrl=%2f&AspxAutoDetectCookieSupport=1

(beef, poultry, or seafood), but it could be a plant-derived ingredient (beans, lentils, cauliflower, etc.). We call this ingredient the *dish star*.

There may be more than a single dish star ingredient in the fridge or pantry and the cook will need to select one. Consider the character of the ingredient and what associations it brings and whether it is in season and at its best. The weather of the day influences whether the choice will be something light (on a hot and humid summer day) or a dish with heavier undertones, such as a stew of beef meat with root vegetables on a cooler or cold day.

Karen Page in her book *The Flavor Bible* (Page & Dornenburg, 2008)[12] has a table that organizes ingredients by characteristics into light, medium, and heavy, and the list of ingredients covers a spectrum from wines and vegetables to grains, fruits, different proteins, and sauces.

The home cook could look at a list of pairings organized by type of ingredient and perhaps cooking method. The pairings may have a strong affinity and enhance each other; some will have a weaker effect; and some types of ingredients may not have a good pairing, e.g., #15 in the example below. This list of pairings could contain the following examples:

1. **Dish star:** e.g., **beef**
2. **Other stars:** pork
3. **Allium or base ingredients:** garlic
4. **Vegetables:** carrots
5. **Complex carb ingredients:** rice, potatoes
6. **Cruciferous vegetables:** broccoli
7. **Herbs:** sweet (basil), savory (bay leaf, thyme, oregano)
8. **Spices:** earthy (cumin)
9. **Umami enhancers:** soy sauce, Worcestershire sauce, mustard
10. **Hot peppers (capsaicin):** peppers from capsicum family (red, amarillo), paprika
11. **Condiments:** lemon juice, mustard, sugar, salt
12. **Other ingredients:** bread, red wine, smoke flavors

---

[12] Karen Page and Andrew Dornenburg, *The Flavor Bible: The Essential Guide to Culinary Creativity, Based on the Wisdom of America's Most Imaginative Chefs* (New York: Little, Brown & Company, 2008, [35.].

13. Fruits: tomatoes
14. Dairy and fats: Butter, olive oil
15. Nuts and seeds: none.
16. Simmering liquids: beef broth

The names of the ingredients that have an extraordinarily strong pairing will be in **bold**. Some groupings may be counterintuitive, such as **#3** allium and base ingredients refer to ingredients in the allium family or used in some regional cuisines as a base for many dishes; in some kitchens, this soffrito or aderezo will be prepared first thing in the morning. Number **5** complex carb ingredients will contain things like pasta. In the pungent group, we will see often ají amarillo, a Peruvian chili with only moderate heat but is very savory and may enhance the umami of many dishes. Once cooked, the heat is tamed. In **#13** – fruits, several are fruits in the botanical sense but are considered vegetables in the culinary sense. I elected to keep them in the fruit group to emphasize the pairing (e.g., tomatoes, avocados, olives).

There are many classical pairings and certainly one should use them because usually there is a good reason why they are paired so often, but you may want to add a little different twist from the classic presentation. Such classic pairings include things like tomatoes and basil, roasted beets with salty cheese, lamb with mint, pork with sage, and some uncommon combinations such as mango with mild chili in a salsa.

To facilitate the process of pairings and to select ingredients already at hand, the following pairings list by dish star may be helpful.

Pairings by dish star ingredient Dish star: beef

1. **Other stars:** pork, **lobster**
2. **Allium and base vegetable: garlic, onion,** shallot
3. **Vegetables: celery, carrots,** green onions, green peas, sweet peppers, mushrooms
4. **Complex carb ingredients: potato**, rice, pasta
5. **Cruciferous vegetables:** broccoli
6. **Herbs:** sweet (basil), savory (bay leaf, thyme, oregano)

7. **Spices:** earthy (cumin)
8. **Umami enhancers:** soy sauce, **Worcestershire sauce**, mustard
9. **Hot Peppers (capsaicin):** hot peppers (ají amarillo, red), paprika
10. **Condiments:** lemon juice, salt, sugar (brown)
11. **Other ingredients:** bread, red wine, smoky flavors
12. **Fruits:** tomatoes, orange (zest, juice)
13. **Dairy and fats: butter**, olive oil
14. **Nuts and seeds:** none.
15. **Simmering and marinating liquids: beef broth**

Creating your recipe: beef daube. Choose the ingredients that go well with beef. **Other protein group** (pork belly); **Allium group** (garlic, onion, shallots); **Vegetable group** (carrots, celery, leek); **Complex carbohydrates group** (to serve with mashed potatoes or rice); **Herbs group** (parsley, rosemary, thyme, sage, bay leaf); **Spices group** (cloves, cumin); **Condiment group** (salt, pepper); **Other ingredients group** (wine, flour); **Fruits group** (orange peel, tomatoes, olives); **Fats group** (olive oil); **simmering liquid group** (marinade)

With all these pairings, then we can go to the recipe, which you can modify to your own taste, maybe adding some pungent ingredient, using a different wine, etc.

Example recipe

Daube (Beef stew Provence style)

Servings: 6

INGREDIENTS

Marinade

- **Aromatics**: 1 carrot, cut into 1-inch sections, 1 onion, quartered, 4 cloves garlic, 1 leek (white part), cut into 3, 1 stalk celery
- **Liquid**: bottle Provence red wine (preferably full-bodied)
- **bouquet garni** (thyme, rosemary, summer savory ground, and bay leaf)
- **Seasonings**: Salt, Pepper, 3 strips orange zest, 3 cloves

Stew

- **Protein**: Two lbs. beef flank (beef stew) sliced, ½ lb. smoked pork belly, diced
- **Aromatics**: 3 carrots, cut into 2-inch sections, 4 shallots, finely chopped, 1 red onion large, finely chopped
- **Fruits: 1** cup black olives, pitted, 6 Roma tomatoes, peeled, seeded, and coarsely chopped
- 2 Tbsp. flour
- 6 tablespoons olive oil
- **Seasoning**s: Salt, Ground pepper to taste.

INSTRUCTIONS

Marinade

1. The evening before, cut the beef into large chunks and place in a large bowl.
2. Add the onion, cut into 4, and insert the 3 cloves. Add the carrot, 2 garlic cloves lightly crushed with the flat side of a knife, and 2 pressed garlic cloves. Add bouquet garni and orange zest. Season with salt and pepper. Cover with red wine. Mix well.
3. Cover with plastic wrap and let stand for at least 8 hours in the refrigerator. Stir the marinade 2 or 3 times during this time.

Stew

1. Drain the pieces of meat with a skimmer and place on paper towels. Reserve the marinade.
2. In a cast iron pot, Dutch oven, or electric slow cooker, heat the olive oil and sweat the shallots and onion over medium low heat.
3. Add the smoked pork belly and sauté for 3 minutes over medium heat. Add the meat and brown the pieces of beef on each side.
4. Pour the flour gradually and stir with a wooden spoon.
5. Add the tomatoes, season with salt and pepper, and mix again.
6. Remove the celery and leek from the marinade and add the marinade to the pot. Cook 1 minute over high heat and simmer over very low heat for 5 to 7 hours or more. At hour 5 you may, if desired, stop the

cooking, allow it to cool, and refrigerate overnight. In the morning you will have a layer of solid fat; remove it, go back to the fridge, and 2 hrs. before serving time, add the carrots and olives, bring to a simmer for 2 more hours, and serve. May use this time to prepare the potatoes, rice, or pasta to accompany this stew.

7. Ensure that the sauce does not completely evaporate during cooking; if so, add some beef broth or water.

Steak Umami Rich sauce

If you found in the refrigerator some nice steaks, maybe flank steaks (tasty and slightly chewy) you may want to have a sauce to go with them and allow your accompanying complex carbs to absorb the sauce. This sauce is rich in umami enhancers, highlights the flavor of the beef, but also allows for the accompanying complex carbs to step up to a higher savory profile.

Servings: 8

INGREDIENTS

- **Liquids**: 1 cup red wine, full bodied, 1 cup of beef broth, 1 Tbsp. of heavy cream, optional
- **Umami enhancer**: fillets of white anchovies, mashed with a fork.
- **Extender**: 6 dried shiitake mushrooms, cut in half, shallot cubed small
- **Seasoning and flavorings**: 1 Tbsp. of brandy, Salt, and pepper to taste

INSTRUCTIONS

1. Soak the dried shitake in the beef broth for 1 hour or more.
2. Prepare the soffrito (aderezo) by sautéing the shallot in 1 Tbsp. of butter until soft. Add the mashed anchovies and cook for a minute or so. Then add the beef broth and mushrooms, cook for a few minutes.
3. Add the wine, the brandy and the cream if using it. Simmer until the mushrooms are soft. Adding salt and pepper to your taste. If necessary, add corn starch slurry (1:1 mix of corn starch and cold water or cold beef broth) to desired thickness.
4. Serve on the meat and complex carbohydrates.

If you wish to apply rule #3 of flavors and provide a contrasting undertone could add a fruit like blueberries or cubed pears.

Steak with creamy mushroom steak sauce with blueberries Steak Umami Rich sauce

Pairings by dish star ingredient Dish star: pork

1. **Other stars:** beef, chicken, **eggs**
2. **Allium or base ingredients: garlic**, ginger, **onion**
3. **Vegetables:** carrots, celery, **green onions**, green beans, mushrooms, peas, **sweet peppers**
4. **Complex carb ingredients:** potato, rice, pasta, bread

136

5. **Cruciferous vegetables: cabbage**
6. **Herbs:** sweet (cilantro), savory (oregano, sage, thyme)
7. **Spices:** earthy (cumin)
8. **Umami enhancers:** mustard, **soy sauce**
9. **Hot peppers (capsaicin):** hot peppers (ají amarillo, red), paprika
10. **Condiments:** ketchup, black pepper, salt, sea salt, vinegar
11. **Other ingredients:** sugar, brown sugar, honey
12. **Fruits**: tomatoes
13. **Dairy and fats:** butter, **olive oil**
14. **Nuts and seeds:** none.
15. **Simmering liquids: chicken broth**

Example recipe

Peruvian style Chicharrón

Chicharrón is a Peruvian staple for breakfast sandwiches on Sunday mornings. In this recipe, the following pairings for pork work well: sweet potatoes, onions, ají amarillo, olive oil, cilantro.

Servings: 6

INGREDIENTS

Chicharrones

Protein: 2 lb. of pork ribs or another fatty cut

Rub: Salt to taste

Liquid: water as necessary.

Onion creole sauce (salsa criolla)
½ red onion finely sliced
1 ají amarillo sliced (either frozen or bottled pickled), or 2 tsp. of ají amarillo paste.
Juice of 2 limes
2 Tbs. olive oil
Salt and pepper to taste

Chopped cilantro leaves to taste.

INSTRUCTIONS

Chicharrones

1. In a cast iron Dutch oven place rinsed rib, add water enough to cover ¼ of the thickness of the meat, add salt and boil until the water is fully evaporated and the meat begins to fry in its own fat (30 minutes +).

2. Begin turning the meat so it turns golden on all sides. Check internal temperature (done at 165°F or 74°C). Once the done temperature is reached remove from Dutch oven, cut in individual serving portions and serve with fried sweet potatoes slices, bread rolls and creole onion sauce.

Example recipe

Onion creole sauce

1. Tame the onion: in a strainer salt the sliced onions abundantly, massage them with your hands for a minute or so, then rinse with hot water to remove most or all the salt. Place them in a bowl.

2. Combine the washed onions with the sliced seedless ají amarillo or the paste, salt, pepper, lime juice and olive oil. Mix gently and sprinkle with the cilantro leaves.

Creole onion sauce(Pickled onions)

Example recipe Roasted Pork tenderloin.

This is an Asian Peruvian fusion recipe, immensely popular, usually served with a contrasting flavor of pickled turnips.

Example recipe

Roasted Pork Tenderloin (CHA SIU)

Ingredients for 4 servings:

- For 2 lb. of Pork Tenderloin
- 2 Tbsp. of sugar
- 2 tsp. of soy sauce
- 4 tsp. of salt

Preparation: Chinese style

- Cut the tenderloin lengthwise in strips of 12" length, 1 ¾ wide and 1/3" thick.
- Place the strips in a cookie sheet with aluminum foil, dust the meat with salt and sugar, sprinkle with the soy sauce, and mix it all so the meat is uniformly covered with the mixture. Let it marinate for a bit.
- Place the meat in a wire baking cooling rack with the cooking sheet bellow, in a preheated oven at 350° F, roast for about ½ hour or internal temp. of 140°F.
- Let the meat rest for 10' before slicing thin.

Example recipe

Pickled Turnip: (Daikon)

- 1 Daikon or purple top turnip peeled and sliced fine in a mandolin.
- Ají Amarillo
- 1 cup of white vinegar
- Sugar 5 Tbsp.
- 1 Tbsp. salt Kosher
- Enough water to cover all the Daikon.

Procedure

1. Wash the daikon slices in salty water multiple times, until the slimy feeling of the daikon or turnip is no longer present (4-5 times) changing the salty water each time.
2. Place the daikon or turnip in a jar layered.
3. Mix the other ingredients in a bowl, boiled for a few minutes, then pour over the sliced daikon.
4. Refrigerate overnight before serving it. Will keep in the refrigerator for up to 2 weeks.

Pairings by dish star ingredient Dish star: **lamb**

1. **Other stars:** none
2. **Allium or base ingredients: garlic**, ginger, **onions**, shallots
3. **Vegetables: carrots**, celery, green onion,
4. **Complex carb ingredients**: potato, pasta
5. **Crucifers Vegetables:**
6. **Herbs:** sweet (cilantro, parsley, mint) savory (bay leaf, oregano, rosemary, thyme, turmeric)
7. **Spices:** fresh (coriander seeds), sweet (cardamom, cinnamon, cloves) earthy (**cumin**)
8. **Umami enhancer:** mustard, soy sauce
9. **Hot peppers (capsaicin):** peppers (ají amarillo, red peppers) cayenne, curry, paprika
10. **Condiments:** lemon, lemon zest, **lemon juice,** red wine vinegar, black pepper, salt (kosher, sea)
11. **Other ingredients: red wine,** white wine
12. **Fruits: tomato**, tomato paste
13. **Dairy and fats: butter, olive oil,** yogurt, milk
14. **Nuts and seeds:** none.
15. **Simmering liquids:** beef broth, **chicken broth**

Example recipe

Lamb chops in seco sauce

In this recipe the lamb chops pair well with chicken broth, beef broth, garlic, onions, rice, parsley, cilantro, cumin, ají amarillo, salt, olive oil.

Servings: 2

INGREDIENTS

- **Liquid for Simmering**: ½ cup of chicken broth, 1 cup beef stock.
- **Aromatics**: 2 Tbsp. chopped flat leaf parsley, ¼ + ¼ cup olive oil, 2+1 Tbs. minced garlic, 1 cup of loosely packed cilantro leaves.
- **Seasoning and Flavorings**: Salt and pepper to taste, ¼ tsp. ground cumin twice
- **Protein**: 4 lamb chops
- ½ cups of cubed medium onions
- 1 Tbsp. ají amarillo paste.
- 1 cup of rice

INSTRUCTIONS

1. In a bowl combine ½ cup of chicken broth, parsley, ¼ cup of olive oil, 2 Tbsp. garlic, ¼ tsp. of cumin and lamb chops. Cover and refrigerate for a minimum of 2 hours.
2. When the marinating is done, heat the oil in a saucepan over medium low heat, add the onions, ají amarillo, and the rest of the cumin. Cook until onions are very soft, about 8-10 minutes.
3. In the meantime, in a blender process the cilantro with some water (1 to 2 Tbsp.) and make a paste. Add it to the beef stock and add the mixture to the saucepan with the onions, bring to a boil, then lower the temp to as low as possible, and simmer for 10 or so minutes to reduce the sauce slightly.

4. Drain the lamb chops from the marinade, dry them in paper towels, and grill them or sauté them, avoid overcooking. The center should be juicy, internal temp 140°F or 68°C.

5. Plate them, pour the sauce over, and serve with rice pilaf style.

Lamb chops seco with rice

Pairings by dish star ingredient Dish star: chicken

Other stars:

1. **bacon, eggs**
2. **Allium or base ingredients: garlic,** ginger, **onions**
3. **Vegetables**: broccoli, carrot, **celery**, **green onions,** lettuce, sweet peppers, squash
4. **Complex carb ingredients:** potato, **rice, pasta**
5. **Cruciferous vegetables:** none
6. **Herbs:** sweet (basil, parsley), savory (thyme)
7. **Spices:** none
8. **Umami enhancer: soy sauce**, Worcestershire sauce
9. **Hot peppers (capsaicin):** hot peppers (ají amarillo, red), curry, paprika
10. **Condiments: BBQ sauce**, salad dressings, lemon juice, lime juice, sugar, salt
11. **Other ingredients**: breadcrumbs

12. **Fruits:** tomato
13. **Dairy and Fats: butter,** milk, cheese (Parmesan), mayonnaise, sour cream, **olive oil**
14. **Nuts and seeds:** almonds, pecans.
15. **Simmering liquids: chicken broth**

Recipe example

Chicken curry

West Indies-style curry with butternut squash and chicken

Servings: 4

In this recipe, chicken pairs well with garlic, ginger, chili pepper (e.g., ají amarillo), onions, squash, rice, curry, lemon juice, oil, milk.

INGREDIENTS

Curry paste

1 onion
1 medium chili
1 garlic clove
2 cm cube ginger, peeled.
2 tsp. ground allspice
1 tsp. fresh thyme
1 tsp. ground fenugreek
½ tsp. ground turmeric
½ tsp. coriander seed
½ tsp. ground cumin
½ tsp. salt

Curry

1 medium butternut squash
1 tsp. coconut oil
14 oz. Chicken breast, diced.

Juice of 1 lemon

7 oz. coconut milk

PROCEDURE

1. Begin by adding all the curry paste ingredients to a food processor and pulse until smooth. I used my trusty Ninja for mine, but you could also use a pestle and mortar.

2. Peel and deseed the butternut squash, cutting it into approximately 1-inch cubes. Set to one side. Easiest to cut both ends off. Set the squash vertically and cut the skin off without wasting the pulp, cut in half lengthwise and remove the seeds.

3. In a large heavy-bottomed pan on a medium heat, add a teaspoon of coconut oil. Once the oil has melted and is hot, add all the curry paste. Fry for 5 minutes, stirring occasionally.

4. While the paste cooks, add the lemon juice to the diced chicken breast, stir to ensure each piece has been coated, then rinse and drain immediately.

5. Add the coconut milk, butternut squash and chicken to the pan, stirring to combine. Put a lid on the pan and leave on a low medium heat in a slow simmer for 30 minutes, stirring every 10 minutes or so.

Chicken Curry with Butternut Squash

6. At this point, the butternut squash should be fork tender and all that is left to do is leave the lid off for 10 minutes for the curry to reduce and

144

thicken. If you prefer hot curry, now is the time to taste it and add an extra chili if you feel it is needed.

Recipe example

Chicken stew

In this recipe, chicken pairs well with chicken broth, garlic, onion, peas, tomato, potatoes, vegetable oil.

INGREDIENTS for 6 servings

For searing

¼ cup vegetable oil

6 chicken thighs and legs

Aromatics

1 ½ medium onions cubed small.

2 garlic cloves, pressed.

2 Bay Leaves

3 tomatoes concasse

Stewing liquid

1 Tbsp. tomato paste

1 cup of chicken broth unsalted

Extenders

6 whole medium Yukon gold potatoes, peeled.

2 cups of green peas

PROCEDURE

1. Sauté the chicken in vegetable oil only until golden, about 7 minutes. Transfer to a plate and keep warm.
2. Prepare the soffrito (aderezo) in the same skillet on medium-low heat; sauté the onion, garlic, and tomato paste, stirring frequently until the onions soften.
3. Put the chicken back in the pan and add the water or the chicken broth and the bay leaves. Cover and lower the heat to low. Simmer until chicken is tender, internal temperature 165°F, about 40 minutes.
4. Add potatoes and cook for about 15 minutes, until potatoes are tender. If using fresh peas, place them in the pot with the potatoes; if frozen, add when the potatoes are done.
5. Serve with rice cooked pilaf method.

Recipe Example

Chicken Escabeche Peruvian Style

Ingredients: for 2 servings

- 2 thighs or breasts of chicken, deboned, skin on.
- Juice of ½ lemon
- ½ Tbsp. of Worcestershire sauce
- 2 red onions, peeled and sliced thick lengthwise.
- ½ Tbsp. minced garlic
- ½ Tbsp. ají Panca paste
- ¼ tsp. grounded cumin
- 1 Bay leaf
- 2 Tbsp. red wine vinegar
- 2 Tbsp. Sherry vinegar
- ¼ cup chicken stock unsalted
- Garnishes with boiled sweet potato ¼ inch slices (to no resistance when pricked with a knife), black olives and farmer cheese cubes.

Procedure:

1. Prepare the chicken skin side down in a sauté pan with oil salt, pepper, sprinkle with lemon juice and Worcestershire sauce. Sauté it first in the stove top for a few minutes in high heat until skin browned, then place the sauté pan in a preheated oven at 500°F for 12 minutes, internal temperature 145°F.

2. While the chicken is in the oven prepare the Escabeche sauce. Boil the onions for 30 seconds to tame them. Remove them and shock them in icy cold water. Empty the saucepan and heat it at medium heat, add the garlic, ají panca, cumin, Bay leaf salt and pepper. Stir for 2 minutes add reserved onions and the vinegar, bring to a boil, and then add the chicken stock, cook to reduce to ½ but keeping the onions crunchy.

3. Plate by placing the sauce over the chicken and arranging the garnishes in the plate.

Chicken escabeche Peruvian Style

Pairings by dish star ingredient Dish star: turkey

1. **Other stars:** eggs
2. **Allium or base ingredients: garlic, onions**
3. **Vegetables: carrots, celery,** green onions, lettuce, mushrooms, **sweet peppers**
4. **Complex carb ingredients:** potato, rice, **pasta, stuffing**

5. **Cruciferous vegetables:** none
6. **Herbs:** sweet (basil, parsley) savory (**bay leaf,** thyme, sage, oregano)
7. **Spices:** earthy (cumin)
8. **Umami enhancer:** soy sauce, Worcestershire sauce
9. **Hot peppers (capsaicin):** chili (amarillo, powder, jalapeño)
10. **Condiments:** lemon juice, sugar
11. **Other ingredients:** breadcrumbs, hamburger bun, stuffing, wine
12. **Fruits:** apple, raspberry, strawberry**,** blackberry, dates dried
13. **Dairy and fats: butter,** cheese (**cheddar**, Monterey Jack, mozzarella, Parmesan, Swiss), mayonnaise, **milk,** sour cream
14. **Nuts and seeds:** none.
15. **Simmering liquids: chicken broth**

Recipe example

Roasted turkey breast

You may dry brine the turkey breast for one to three days prior to cooking or season it at cooking time as in the recipe.

When I was a teenager there was a place next to a Movie Theater that prepared Roasted turkey Sandwiches, so good that sometimes I and my friends would go to the movie theater so we could have the roasted turkey sandwich.

In this recipe, turkey pairs well with lettuce (garnish sandwiches), thyme, cumin, salt, black pepper, onions, white wine, butter, extra virgin olive oil (EVOO).

INGREDIENTS

3-7 lb. thawed skin-on turkey, breastbone in
1 Tbsp. unsalted butter or EVOO
1 Tbsp. kosher salt
½ tsp. black pepper
3 sprigs of thyme
½ tsp. ground cumin

Bread rolls

INSTRUCTIONS

1. Place the turkey breast in the rack and roasting pan.
2. Preheat oven to 450°F or 230°C.
3. Rub the turkey breast with butter mixed with the seasonings, over the skin and under.
4. Place the turkey breast in the rack and roasting pan and into the oven.
5. Lower the heat immediately to 350°F or 175°C and roast for one hour.
6. Check internal temperature (done at 165°F or 74°C. If not done, check the temp every 10 minutes until desired temperature is reached).
7. Check the color of the skin as well when you check the temperature; if it is browning too fast, cover with a strip of aluminum foil.
8. When done, take out of the oven and rest it for 15-20 minutes covered loosely with aluminum foil. Prepare the gravy from the drippings in the roasting pan.
9. When the resting time is up, carve the bone from the meat, staying as close to the bone as possible. Then cut the breast into slices crosswise.
10. Plate as sandwiches, with gravy and creole onion sauce (see recipe example above).

Gravy

INGREDIENTS

Defatted turkey drippings + enough chicken stock for 2 cups heated before using it.

1 stick unsalted butter

1 ½ cups of cubed yellow onions

¼ cup all-purpose flower

1 tsp. kosher salt

½ tsp. black pepper

1 Tbsp. brandy

2 Tbsp. white wine

INSTRUCTIONS

1. In a large saucepan melt the butter on medium low heat, then cook the onions slowly until the onions begin to brown.

Dissolve the flour in cold chicken stock to make a slurry and add it to the saucepan in small portions. Add the hot chicken broth mixture, the brandy, and wine. If the sauce is too thin, add some more slurry until it doesn't thicken any more. Season to taste.

Roasted Turkey Breast with Jicama toast

Pairings by dish star ingredient Dish star: fish

1. **Other stars:** clams, mussels, sea scallops, **shrimp,** eggs
2. **Allium or base ingredients: garlic,** ginger, **onions,** shallot
3. **Vegetables: carrots, celery,** green onions, mushrooms, peas, **sweet peppers**
4. **Complex carb ingredients: potato,** rice, pasta
5. **Cruciferous vegetables: cabbage**
6. **Herbs:** sweet (basil, cilantro, **parsley**) savory (bay leaf, thyme)
7. **Spices:** earthy (saffron), fresh (coriander seeds)
8. **Umami enhancer:** anchovies or anchovy paste, soy sauce
9. **Hot peppers (capsaicin):** cayenne, curry, hot peppers (ají amarillo, red), hot sauce, paprika
10. **Condiments: lemon and juice**, **limes and juice**, kosher, and sea salt
11. **Other ingredients:** sharp spices (pepper black or white), sugar, white wine
12. **Fruits**: passion fruit, pears, pineapple, **tomato**
13. **Dairy and fats: butter,** heavy cream, milk, **olive oil**
14. **Nuts and seeds:** none.
15. **Simmering liquids: fish broth, shrimp broth**

Recipe example

Ceviche general guidelines.

Servings: 4 to 6

INGREDIENTS

Fish, the freshest possible, raw. Any type of fish will work, the best tasting fish is white meat, firm and not oily. Sea bass, flounder, shark, cobia, porgy, African pompano, trigger fish, and bonito are good examples of the kinds of fish that match well with the preparation of ceviche.

Citrus juice: the most common citrus juice used is lime juice. Lemon and limes were brought by the Spaniards when they colonized Peru. Prior to that, ceviche was prepared using local sour fruits like passion fruit and sour oranges.

Ají is a chili pepper (capsicum); the most popular are ají amarillo and ají limo (like the habanero chili)

Onion: the most common used is red onion, sliced thinly, salted liberally for a minute or two, massaged with your hands during that time and rinsed with hot water to remove all the salt.

INSTRUCTIONS

1. The meat of the fish is denatured by a chemical process of the acid (citric). The firmness of the fish meat, the acidity of the lime, and marinating time are the variables that determine how "well cooked" is the fish at the time is served.
2. As you prepare the dish, is important to determine the firmness of the fish, taste the lime juice to determine the acidity, and estimate the time required to obtain the result that is pleasurable for you.
3. The other variable you need to judge is the degree of spiciness of the dish. Ají amarillo is moderately hot at about 40,000 to 50,000 Scoville heat units with a savory flavor; the ají limo or habanero is at about 50,000 Scoville heat units and has a citrusy flavor and aroma.

4. The usual times I would use for a fish that is moderately firm with an average acidic lime is 3-5 minutes from arranging the plate to serving. The amount of chili I would use for ají amarillo paste is 2 tsp. for 3 limes plus 1 lemon juice base sauce. For the same amount of sauce, I would use 1 small ají limo or habanero cut in half lengthwise, remove all seeds and ribs, soak in the ceviche sauce while preparing it and then remove it. Add at this point the onions (tamed – see instructions above in Chapter 6) complete the dish.
5. Ceviche also includes other ingredients, such as a segment of corn on the cob and boiled sweet potatoes.

Ceviche notice fish cut in cubes.

Recipe example

Peruvian style ceviche

In this recipe the fish pairs well with garlic, red onions, corn, lettuce, sweet potato, ají limo or habanero, ají amarillo, lime juice, clam juice.

Servings: 4 to 6

INGREDIENTS

- **Leche de tigre (ceviche sauce)**

2/3 cup fresh lime juice

2 garlic cloves, smashed or pressed.

1 Tbsp. (packed) chopped fresh cilantro leaves.

1/2 ají limo or habanero chili, seeded, halved lengthwise, or 2 tsp. ají amarillo paste or both.

1/2 small red onion, chopped.

1/2 cup bottled clam juice (optional)

Kosher salt

- **Ceviche dish(fish)**

1 small, sweet potato (about 8 oz.)

1 ear of corn, husked.

1/2 ají limo or habanero chili, seeded, halved lengthwise, or 2 tsp. of ají amarillo paste, available at most Latin markets.

1 pound fluke, flounder, sole, or other white firm fish (black seabass, trigger, porgy, and African pompano) cut into 1/2-inch cubes.

1 small red onion, quartered and thinly sliced, divided.

Kosher salt

1 Tbsp. cilantro leaves chopped.

INSTRUCTIONS

Leche de tigre (ceviche sauce)

Set a fine-mesh sieve over a small bowl. Purée the first 4 ingredients and 4 large ice cubes in a blender until smooth. Add onion; pulse 3-4 times. Strain liquid into a medium bowl. Stir in clam juice, if desired, season with salt. Cover and chill.

Ceviche dish (fish)

1. Pour water into a large pot fitted with a steamer basket to a depth of 1 inch; bring to a boil. Add sweet potato, cover, and cook until just fork-tender, about 30 minutes. Transfer to a plate; let cool.

2. Meanwhile, add more water to the same pot, if needed, to measure 1 inch; bring to a boil. Add ear of corn to the pot and steam until crisp-tender, 2-3 minutes. Transfer to a plate; let cool completely.

3. Halve sweet potato lengthwise. Using a chef knife, cut ½" sections, place them in a small bowl, and set aside. Cut kernels from cob. Reserve 1/3 cup kernels (save extra kernels for another use).

4. Rub a large bowl with cut sides of chili; discard, or alternatively add the ají amarillo. Place fish, 2/3 of the onion, leche de Tigre, and 4 large ice cubes in bowl; stir well. Let marinate for 2 minutes; remove ice. Fold in potato and corn; season with salt.

5. Using a slotted spoon, divide ceviche into small bowls or onto plates. Drizzle ceviche with leche de Tigre from bowl; garnish with remaining onion and cilantro.

Recipe example

Tiradito of Corvina

In this recipe, fish pairs well with garlic, corn, lettuce, cilantro leaves, sweet potato, salt, and pepper, ají limo or habanero, ají amarillo, lime juice, clam juice, olive oil, soy sauce. This dish is as well marinated using lime juice, differs from ceviche in that the cut of the fish is slanted, does not use onions, the marinating liquid may contain in addition to lime juice; soy sauce, ginger, and olive oil. Is likely known in many sea food restaurants in America as Crudo.

INGREDIENTS

*   200 grams of fresh corvina, although you can also use other, similar fresh fish, (see ceviche recipe) sliced in fine slanted slices.
*   8 yellow ajíes, without seeds or veins, for the sauce or alternatively 2 Tbsp. of ají amarillo paste.

- 1 Tbs. soy sauce
- 1 tsp. garlic, minced.
- 1 Tbsp. olive oil
- ½ tsp. salt
- 1 pinch of monosodium glutamate (ají no moto)
- Ají limo to taste, cut into large pieces, preferably red- or yellow-colored limo peppers (habaneros) to soak in the juice for a few minutes, then discard.
- 2 ice cubes
- 5 limes
- Cilantro leaves for garnish.
- Kernels of corn
- Cooked sweet potato.

INSTRUCTIONS

To prepare this dish, use very fresh corvina or a similar fish. The fish is best when fresh caught. You should never make skeletal tiraditos (stretched) of fish, that is, slices of smashed fish that has been poorly treated and that does not even attain 3½ oz.

1. Obtain a thick, large fillet of corvina. Wash it very delicately and cut 200 grams (5 oz.) of slices on a cutting board. The cut should neither be too thick nor too thin. The slices should be some 6 cm long. To avoid the fish sticking to the knife, cut the fish while moist and from time to time moisten the edge of the knife. Some people stretch the fish, giving it a blow with the side of the knife blade. If you wish to taste the original texture of the fish when you bite into it, avoid doing this.
2. Place the fish slices, now chilled, on a large plate. Keep it chilled for a few minutes.
3. Make your ají paste by boiling the pods of yellow ají for some ten minutes. Drain and peel back the skin, remove seeds and veins. Blend with a small amount of water. The result should be a thick and creamy sauce. To guarantee the best flavor, do not prepare this more than three

hours early. It can be purchased prepared and bottled at Latin food markets.

4. In a bowl, add the ends and leftover bits of fish and season them with half a tsp. of salt and a pinch of monosodium glutamate. Add 1½ tablespoon of ají paste and two ice cubes and squeeze in the juice of five juicy limes. Soak the thickly diced ají limo for a few minutes; it is best to use the ones colored yellow or red since they give the best scent. Also, do not use the seeds since they will make your dish too spicy to enjoy. Remove the ice and sieve the sauce over the well-placed slices of fish.

5. Decorate with whole cilantro leaves. Serve immediately with lettuce, corn, and sweet potato.

Tiradito covered with sauce.

Recipe example

Sudado de pescado (Poeling, fish stew Peruvian style)

In this recipe fish pairs well with fish broth, chicken broth, garlic, onion, cilantro leaves, ají panca paste, ají amarillo paste, lemon, and tomatoes concasse.

Servings: 2 to 4

Ingredients

One small fish (whole) or two fillets

Aromatics

2 Tbsp. canola oil

2 onions sliced medium thin.

½ Tbsp. garlic pressed

1 Tbsp. ají panca paste.

1 Tbsp. ají amarillo

Salt and pepper to taste

2 Tbs. chopped cilantro.

½ lemon

Liquid to braise.

½ cup chicha de Jora (maize brew) or ½ cup fish or chicken broth

3 tomatoes in concasse (peeled and seeded or Marzano, canned)

INSTRUCTIONS

1. Season the fish with salt and pepper.
2. Prepare the aderezo: in a skillet heat the oil, lower the heat to medium low. You are making layers in the pot: first layer with Ajíes, then one-half of the onion and one-half of the tomatoes.
3. You may add some green onions and cook until the tomatoes have dissolved.
4. Add the simmering liquid (chicha de   Jora - fermented drink from corn), fish or chicken broth); add the fish and cover the fish with the remaining onions and tomatoes.
5. Cover with a tight lid, lower the heat some more, and let it sweat until the fish reaches an internal temperature of 165°F, 74°C, in about 50 minutes.
6. Serve with rice or boiled yucca.

Recipe example

Fish Escabeche

This is a dish brought to Peru by the Spanish during the Colony, the name escabeche has an origin in the Arab word *sikbaj pronounced iskebech.*

Ingredients: for four servings

- ¼ cup red wine vinegar.
- 4 medium sized boiled sweet potatoes.
- 2 hard boiled sliced eggs
- 8 Alfonso olives halved.
- 4 small red onions cut into thick slices lengthwise.
- 2 medium yellow chilies sliced, seeds removed or ají amarillo paste   1 ½ tsp.
- 1 tsp. of ají panca paste.
- 1 cooked corn on the cob.
- 1 tsp. mustard prepared.
- 1 tsp. crushed garlic.
- 4 Tbsp. vegetable oil for frying.
- ½ cup of all-purpose flour.
- Salt, pepper, cumin, and oregano to season
- Four filets of white fleshed fish (grouper, mahi-mahi, corvina etc.)

Procedure:

1. Season fish fillets with salt and pepper then flour them on both sides.
2. Pre-heat a frying pan, add 2 Tbsp. of frying oil and fry the fillets in medium heat until golden 10-15 minutes, turning them halfway and to internal temperature (135°F or 57 ½ °C) will reach safe temperatures while it rests.
3. In another frying pan add the remaining oil and start sautéing the onions, the yellow and panca chilies pastes, garlic, pepper, and cumin. Continue to fry for a minute or two, add the vinegar and then some oregano.
4. Turn off the heat and plate with the fish as center piece, douse generously with the sauce, add 1-2 slices of hard boil egg, an olive or two, a segment of corn on a cob, and a thick slice of sweet potatoes.

Grouper Escabeche

Pairings by dish star ingredient Dish star: shrimp

1. **Other stars**: chicken, crabmeat, eggs, sea scallops
2. **Allium or base ingredients: garlic,** ginger, **onions**
3. **Vegetables:** carrots, **celery, green onions,** mushrooms, peas, **sweet peppers, fennel.**
4. **Complex carb ingredients: potato, rice, pasta**
5. **Cruciferous vegetables: cabbage**

6. **Herbs:** sweet (basil, cilantro, **parsley**) savory (oregano, thyme)
7. **Spices:** fresh (anise)
8. **Umami enhancer:** soy sauce, Worcestershire sauce
9. **Hot peppers (capsaicin):** cayenne, peppers (**ají amarillo**, red), paprika
10. **Condiments:** lemon, **lemon juice,** lime, lime juice, salt, black and white pepper
11. **Other ingredients:** breadcrumbs, club soda, honey, caramel, **white wine**
12. **Fruits: tomatoes, Avocado.**
13. **Dairy and fats: butter,** heavy cream, milk, mayonnaise, **olive oil,** Parmesan cheese
14. **Nuts and seeds:**
15. **Simmering liquids: chicken, shrimp, or fish broth**

Recipe example

Honey garlic shrimp with ají Amarillo

In this recipe, shrimp pairs well with garlic, ginger, scallions, broccoli, asparagus, green beans, rice, soy sauce, honey, ají Amarillo or red.

INGREDIENTS

Marinade

Protein

- lb. shrimp, peeled.

Aromatics

1½ Tbsp. garlic, minced.

- tsp. ginger, minced.
- scallion, chopped.

Marinade

- 1/3 cup honey
- 1/3 cup soy sauce
- 1½ tsp. Ají Amarillo paste

Vegetables

- Broccoli, asparagus, or Green beans cooked. To reheat with the shrimp and the marinade

For sautéing:

1 Tbsp. EVOO

INSTRUCTIONS

1. In a plastic bag, mix the shrimp and the marinade. Refrigerate for 15-30 minutes.
2. In a sauté pan, heat the oil, sauté the marinated shrimp, and then add marinade liquid.
3. Add the vegetables of your choice.
4. Serve with rice and vegetables (broccoli, asparagus, or green beans).
5. Pour sauce overall.

Shrimp with Honey, Soy Sauce, Ají Amarillo, Ginger, and lentils

Pairings by dish star ingredient Dish star: crabs

1. **Other stars:** chicken, **clams, crawfish**, **eggs,** fish, lobster, mussels, octopus, **oysters, sea scallops, shrimp,** sausage (pork)
2. **Allium or base ingredients: garlic, ginger, onions, shallots**
3. **Vegetables:** carrots, **celery**, corn, fennel, **green onions**, leek, lettuce, **okra**, salad greens, spinach, **sweet peppers**
4. **Complex carb ingredients: potato, rice, pasta**
5. **Cruciferous vegetables:** none
6. **Herbs:** sweet (**basil, cilantro, parsley**, tarragon) savory (bay leaf, chives, **thyme, oregano**)
7. **Spices:** sweet (cloves), fresh (coriander seeds), sharp (black and white pepper), earthy (cumin, saffron)
8. **Umami enhancer:** mustard, oyster sauce, **soy sauce**, Worcestershire sauce
9. **Hot peppers (capsaicin): cayenne, hot peppers** (ají amarillo, red, green), hot sauce, paprika, mustard
10. **Condiments: old bay seasoning, lemon, lemon juice, lime, lime juice, salt** (kosher, sea, table)
11. **Other ingredients:** breadcrumbs, bread, **corn meal**, sugar, **white wine**, beer, club soda
12. **Fruits: avocado, tomato, and paste**
13. **Dairy and fats:** bacon grease, coconut milk, **olive oil,** milk, mayonnaise, heavy cream
14. **Nuts and seeds:** none.
15. **Simmering liquids:** chicken broth, shrimp broth

Example recipe

Crab meat soufflé

In this recipe the crab meat pairs well with eggs, milk, heavy cream, salt, chili, lime.

Servings: 6

INGREDIENTS

- 3 tablespoons butter
- 3 tablespoons all-purpose flour
- 1 ½ cups of half-and-half
- 3 eggs, separated + 1 egg white.
- 3/4 tsp. dry mustard
- 1/2 tsp. salt
- 1/8 tsp. red pepper powder
- 1-pound lump or flake crab meat, drained
- 1 tsp. lime juice
- 1/4 tsp. cream of tartar
- 1 scallion minced.

INSTRUCTIONS

Preheat the oven to 325°F, set the rack low.

1. Melt butter in a large heavy saucepan over low heat; add flour, stirring until smooth. Cook one minute, or until a nutty aroma develops, stirring constantly. Gradually add half-and-half; cook over medium heat, stirring constantly, until thickened and bubbly. Remove from heat. Set it aside.

2. Beat egg yolks until thick and paled to a lemon color. Gradually stir one-fourth of hot white sauce mixture into yolks to temper them, mix well, then add to remaining hot white sauce, stirring well. Stir in mustard, salt, and pepper.

3. Combine crab meat and lime juice; add the scallion, stir crab meat mixture into white sauce.

4. Beat egg whites (at room temperature) until foamy; add cream of tartar, beating until egg whites are stiff but not dry. Gently fold into it the cooled down crab meat mixture.

5. Butter the bottom of a 1 1/2-quart soufflé dish; pour the crab meat and egg white mixture into dish. Place dish in a large pan; add hot water to pan to a depth of 1 inch.
6. Bake at 325° for 1 hour and 15 minutes or until puffed and golden brown. Serve hot.

Pairings by dish star ingredient Dish star: squid

1. **Other stars:** chicken, **clams, eggs,** fish (cod, seabass, **monk fish, grouper**), lobster, **mussels**, octopus, **scallops, shrimp**, oysters, sausage (pork)
2. **Allium or base ingredients: garlic, ginger, onions, shallots**
3. **Vegetables:** bean sprouts, carrots, leeks, peas, celery, cucumbers, fennel, **green onion, sweet peppers**, mushrooms, salad greens, spinach, **zucchini**
4. **Complex carb ingredients:** potato, **rice, pasta**
5. **Cruciferous vegetables:** none
6. **Herbs:** sweet (**basil**, cilantro, mint, **parsley**) savory (**bay leaf, oregano**)
7. **Spices:** sharp (**black peppercorn**) earthy (saffron, **thyme**)
8. **Umami enhancer: soy sauce**
9. **Hot peppers (capsaicin):** cayenne, **chili (ají amarillo, red)**, paprika
10. **Condiments:** clam juice, **lemon**, lime, **lemon** juice, lime juice, lemongrass **fish sauce**, oyster sauce, white and red vinegar, rice, salt (**kosher, sea**), vinegar (**balsamic**), **sherry** vinegar.
11. **Other ingredients:** bread, breadcrumbs, sugar, red wine, **white wine**
12. **Fruits: tomatoes** and paste, olives
13. **Dairy and fats: Butter,** milk, **olive oil**
14. **Nuts and seeds:** almonds, peanuts, sesame seeds, **walnuts.**
15. **Simmering liquids:** chicken stock, clam juice, **fish stock**

Recipe example

Calamari salad

In this recipe squid pairs well with garlic, onions, celery, flat leaves parsley, lemon juice, vinegar, salt and black pepper, tomatoes, olives

Servings: 4 main course servings

INGREDIENTS

1-1/2 lb. cleaned baby squid.
2 tablespoons fresh lemon juice
1 Tbsp. red wine vinegar
1/3 cup extra virgin olive oil.
1 large garlic clove, minced.
1/2 tsp. salt
1/4 tsp. black pepper
1 small red onion, halved lengthwise, then thinly sliced (1 cup).
1/3 cup pitted Kalamata or Peruvian Botija olives, halved lengthwise.
2 cups cherry or grape tomatoes (3/4 lb.), halved or quartered if large.
2 celery ribs, peeled, cut into ¼ -inch thick slices lengthwise, cubed small.
1 cup loosely packed fresh flat-leaf parsley leaves.

INSTRUCTIONS

1. Rinse squid under cold running water, then lightly pat dry between paper towels.
2. Halve tentacles lengthwise and cut bodies (including flaps, if attached) crosswise into 1/3-inch-wide rings, if small baby calamari may use whole.
3. Cook squid in a 5- to 6-quart pot of boiling salted water, uncovered, until just opaque, 40 to 60 seconds.
4. Drain in a colander and immediately transfer to a bowl of ice and cold water to stop cooking. When squid is cool, drain and pat dry.

5. Whisk together lemon juice, vinegar, oil, garlic, salt, and pepper in a small bowl, then stir in the onion and let stand 5 minutes.

6. To complete combine squid, olives, tomatoes, celery, and parsley in a large bowl. Toss with the dressing and season with salt and pepper. Let stand at least 15 minutes to allow flavors to develop.

Recipe example

Deep fried squid with anchovy-laced grits

Ingredients

Grits

8 slices bacon, cut into 1/2-inch pieces.
1 yellow onion, chopped.
2 cups stone-ground grits (coarse the better!)
4 cups low-sodium chicken broth
6 anchovy fillets
2 cups heavy cream
Dash of cayenne or 2 tsp. ají amarillo paste.
1 1/2 cups grated Monterey Jack or Parmesan cheese.
Salt and freshly ground black pepper.

Deep-fried squid

Ingredients:

2 Tbsp. butter
2 Tbsp. olive oil
Salt and freshly ground black pepper.
2 Tbsp. apple cider vinegar
2 Tbsp. honey
1 Tbsp. cinnamon

PROCEDURE

Grits

1. In a large pot over medium heat, cook the bacon and onions until the bacon is chewy and the onions are translucent, about 1 minute.

2. Add the anchovies, mashed with a fork, and cook for a minute or so.

3. Pour in the grits and add the chicken broth. Stir together and bring to a boil. Reduce to a simmer, cover, and cook, stirring occasionally, for 30 to 40 minutes.

4. Add the cream and ají amarillo, stirring them into the grits.

5. Cover the pot and keep cooking over very low heat, stirring occasionally to keep them from sticking, until the grits are tender but still have a bite to them, another 20 to 30 minutes.

6. Just before serving, stir in the cheese. Add salt and pepper to taste.

Squid

1. Marinate the squid in all the other ingredients mixed in a Ziplock plastic bag and refrigerate for at least   30 minutes or longer. Probably should be the first step (before making the grits).

2. Scratch the surface of the body of the squid in a crisscross pattern of relatively small diamonds.

3. In a deep fryer at high temperature, near smoking temp for the oil being used, fry the squid for a minute or so, and remove as soon as it starts turning golden.

4. To serve, spoon a generous helping of grits onto a plate. Lay 4-5 squid's bodies on top of the grits.

5. Cook the marinade to a boil and thicken it with corn starch slurry if necessary and use as a sauce on the grits.

Servings: 6 to 7 cups

Example Recipe

Stuffed Calamari over Pasta

Ingredients: for 2 servings

FOR THE SQUID FILLING: should be some left over to add to the sauce.

- 16 medium size or 12 large size squid with tentacles if possible.
- 6 bread sticks or ½ cup + 2 Tbsp. of Panko.
- 2 Tbsp. of olive oil light.
- ¾ Tbsp. unsalted butter.
- 2 ½ garlic cloves mushed, skin off.
- 1 espresso coffee spoon of Ají amarillo paste.
- 16 peeled shrimp cut in thirds or fourths depending on size.
- Kosher salt
- ¼ cup of dry white wine (Sauvignon Blanc)
- Lime zest ¼ tsp. about zest of ½ a lime.

SAUCE: is basically a tomato sauce enhanced by the tentacles and the leftover of the filling of the squid.

- 2 Tbsp. of canola oil or olive oil light.
- Ají amarillo paste.
- 2 garlic cloves mushed, skin off.
- 1 can Tomatoes Marzano type chopped.
- Kosher salt and fresh ground black pepper
- 2 Tbsp. Parsley or cilantro leaves chopped.

Procedure:

For the stuffed squid.

1. Prepare your ingredients. Crush the bread sticks in a resalable plastic bag, to coarse crumbs. Cut the tentacles of the squid and chop them finely. If you could not get the tentacles get some extra bodies and mince them to about 1mm size (smaller than 1/16 of an inch). Cut the shrimp to the appropriate size for the size of the squid you are using.

2. Prepare the soffrito or aderezo for the filling. In a frying pan heat 2Tbsp. of olive oil, keep the heat low, with soffrito always use low heat. Add the tentacles or minced bodies and cook them to golden brown. Then add the aromatics, butter, garlic and ají amarillo. Cook low and slow until garlic begins to turn golden (maybe a minute or so). Season with salt to taste.
3. Add the protein (shrimp) and cook to pink, add the wine and simmer just enough to evaporate the alcohol. All these steps occur rapidly, a minute or two at the most for each.
4. Complete the filling by adding the bread sticks crumbs or Panko and the lime zest. The product should look like wet sand. If it is too dry add some simmering liquid (wine). Turn the heat off, move the pan to let it cool to warm so you can handle it.
5. Fill the squid bodies with the filling, works best using your hands and small spoon. Do not overfill them, it will expand when cooking in the sauce. Seal the top of the squid bodies with toothpicks.
6. At this point may refrigerate them if you prepare them early.

For the sauce:

1. Prepare the soffrito for the sauce. In a deep wide pan heat the oil, add the aromatics garlic, ají amarillo paste and if you wish could add some shallots or onions minced small. Cook until fragrant and just starting to color.
2. Add the tomatoes with the liquid from the can. Season with Salt and pepper lightly because the juices will reduce increasing the concentration of the seasoning. Bring to a boil and then reduce the heat to keep a gentle simmer for 15-20 minutes.
3. Add the stuffed calamari and the leftovers of the filling, bring back to a simmer. Cover the pan and keep the gentle simmer for 45 minutes or so.
4. Adjust the seasoning. Serve with Pasta, or alone as an appetizer.

Pairings by dish star ingredient Dish star: Octopus

I have heard that the type of octopus best suited for the kitchen is the rock (rocky habitat) octopus that is readily recognizable because it has two rows of suckers in the tentacles. Whether this is true, I am not sure. I have searched for a reliable source and have not found it.

1. **Other stars: clams**, crabs, eggs, fish, **lobster**, **mussels**, **scallops, shrimp, squid,** tuna.
2. **Allium or base ingredients: garlic,** ginger, onions, **shallots**
3. **Vegetables:** arugula, asparagus, broccoli, **cannellini beans, carrots, celery,** corn, cucumber, chickpeas, fennel, **green onions,** lima beans, **mushrooms, sweet peppers**
4. **Complex carb ingredients:** plantains, **potatoes, rice, pasta**
5. **Cruciferous vegetables:** none
6. **Herbs:** sweet (**basil**, cilantro, marjoram, mint, **parsley**) savory (**bay leaf,** chives, **oregano,** rosemary, **thyme**)
7. **Spices:** Sharp (peppercorns) Sweet (cloves) Earthy (**cumin**)
8. **Umami enhancer:** anchovy filet, bonito flakes, **caviar**, mustard, Tapenade, shiitake mushrooms, nori, soy sauce, Worcestershire sauce
9. **Hot peppers (capsaicin):** cayenne, peppers (ají amarillo, red, green), wasabi
10. **Condiments:** balsamic vinegar, brine, **lemon, lemon juice** and zest, lime juice, red and white wine, sea salt
11. **Other ingredients:** Almonds, pernod, wines (red, rice, white), sesame seeds
12. **Fruits: tomatoes** and sauce and paste, orange
13. **Dairy and fats:** butter, mayonnaise, **olive oil**
14. **Nuts and seeds: almonds.**
15. **Simmering liquid:** fish stock

Recipe example

Rock octopus braised and grilled

In this recipe the octopus' pairs well with garlic, onions, potatoes, bay leaf, parsley flat leaves, pimentón (Spanish smoked sweet paprika), salt, white wine.

Check out this video[13] for a Spanish-Style Braised Octopus[14]. Much helpful information is also available on the Food Wishes site, for a paid membership, not associated to us.

INGREDIENTS

Octopus

1 lb. of rock octopus
2 Tbs. olive oil
1 sweet yellow onion
3 cloves of garlic
1 large bay leaf
1 ½ tsp. pimentón
1 ½ tsp. kosher salt
½ cup white wine

Sauce

1/3 cup reserved braising liquid, boiled and strained.
1 Tbsp. fresh lemon juice
1 Tbsp. EVOO
Salt to taste

---

[13] YouTube, Spanish Octopus - Spanish-Style Braised Octopus Recipe, accessed date June 23 2021 https://www.youtube.com/watch?v=onCfFc8FNr8

INSTRUCTIONS

1. Prepare the soffrito (see the section about base preparations - soffrito, aderezo, mire poi, etc. in Chapter 7) with onions, garlic, bay leaf, pimentón, EVOO, and salt. Set the heat at medium and cook with occasional stirring until the onions soften. Add the wine and wait for it to begin to boil.

2. Then add the octopus (whole tentacles) and cover it with the soffrito by turning it a few times.

3. Turn the heat to low, cover, and simmer for about an hour, then check for tenderness with a paring knife tip. When it reaches the tenderness, you like (don't let it get mushy) remove from heat, and place in a bowl to cool with the braising liquid.

4. Once cold, refrigerate overnight covered in plastic.

5. When ready to serve, remove the octopus, wipe with paper towels to remove excess braising liquid and debris of onions, etc. Cut it in 2-3 sections to make handling easy.

6. Heat the grill, and in the meantime heat the braising mixture to a boil and then strain it.

7. Grill the 2-3 segments of octopus on high to caramelize the surfaces, turning it to cover all surfaces.

8. Douse it with EVOO and let it rest.

9. Prepare the sauce with the boiled and strained braising liquid, and add lemon juice, EVOO, and parsley.

10. Plate and serve with crusty roasted potatoes.

Octopus simmered and grilled

Recipe example: José Andrés'[15] Pulpo a la Gallega (grilled octopus with potatoes)

Servings: 4

José Andrés adds a clean copper penny to the boiling pot to replicate the traditional technique of cooking octopus in a copper pot. Soak the penny in distilled white vinegar and salt for 10 minutes, then scrub it clean before using. The chemical reaction of the copper and the octopus gives the tentacles a gorgeous purple tint.

Octopus Grilled with potatoes Galicia Style (Pulpo a la Gallea)

Inspired by José Andrés

[15] Food & Wine, Pulpo a la Gallega, accessed date June 23, 2021
https://www.foodandwine.com/recipes/pulpo-la-gallega-grilled-octopus-potatoes

INGREDIENTS

For First Boil

¼ cup kosher salt

1 Tbsp. of black peppercorns

1 bay leaf

One 5½ lb. octopus, cleaned. Should be a rock octopus (2 lines of suckers per tentacle)

1 clean penny if using

For second boil

1¼ lb. Yukon gold potatoes, peeled.

For grilling

4 Tbsp. extra virgin olive oil

Finishing and serve

1 Tbsp. lemon juice

¼ tsp. of pimenton (smoked sweet paprika)

Sea salt, flaky

Parsley leaves for garnishing

INSTRUCTIONS

1. **First boil.** Boil 8 quarts of water in an adequate-size pot; add the rest of the ingredients for first boil except the octopus. Dip the tentacles in the boiling water 3 times, then lower it into the pot. Reduce the heat to moderately low, simmer for 75 minutes, weight the octopus so it is submerged - a plate inversed works well, or a smaller pot lid.

2. **Second boil.** Add the potatoes and cook them in a simmer for about 25 minutes; remove the octopus, cut the tentacles, and wipe the purple skin

from them with paper towels. Slice the potatoes in 1/8-inch slices and discard the braising liquid and head of the octopus.

3. **Grilling.** In a grill or a hot grilling pan grill the tentacles after brushing them with 3 Tbsp. of olive oil, turning them to grill all surfaces to a light char, about 5 minutes.

4. **Finish**. Arrange on a plate an octopus tentacle with the potatoes, drizzle with lemon or lime juice, and a Tbsp. of olive oil. Sprinkle with sea salt, and garnish with parsley leaves. pimentón,

Pairings by dish star ingredient Dish star: scallops

1. **Other stars:** bacon, chicken, clams and juice, lobster, mussels, shrimp, sausage (pork)
2. **Allium or base ingredients: garlic,** ginger, **onions, shallot**
3. **Vegetables:** carrots, celery, **green onions,** mushrooms, peas, sweet peppers
4. **Complex carb ingredients:** potato, rice, **pasta**
5. **Cruciferous vegetables:**
6. **Herbs:** sweet (basil, cilantro) savory (thyme)
7. **Spices:** earthy (saffron)
8. **Umami enhancer:** Dijon mustard, soy sauce
9. **Hot peppers (capsaicin):** cayenne, hot peppers (**ají amarillo, red, green**), paprika
10. **Condiments:** kosher salt, **lemon, lemon juice,** lime, lime juice
11. **Other ingredients:** bread, breadcrumbs, sugar, **white wine**
12. **Fruits: tomatoes**
13. **Dairy and fats: butter, cheese (Parmesan) heavy cream**, milk, **olive oil**
14. **Nuts and seeds:** none.
15. **Simmering liquids**: chicken broth/stock

Recipe example

Conchitas a la Parmesana (Parmesan scallops)

In this recipe the scallops pair well with lime juice, Worcestershire sauce, salt and black pepper, butter, and Parmesan cheese.

Servings: 2

INGREDIENTS

12 bay scallops in a half shell (available most often frozen and only the round muscle. May place them in ceramic scallop shells or in Chinese soup spoons)
Salt and pepper
12 drops Worcestershire sauce
12 drops lime juice
4 tablespoons butter
12 tablespoons grated Parmesan cheese.

INSTRUCTIONS

1. Preheat the broiler on high.

2. Clean and wash the scallops in the half shell, dry and season with salt and pepper (if the shells not available use Chinese soup spoons).

3. Put them in a cookie tray and season each one with a drop of Worcestershire sauce and one to two drops of lime juice. Cover with one tablespoon grated cheese and ¼ teaspoon butter.

4. Run under the broiler for a few minutes (about 4) or until the cheese is bubbling and golden brown.

5. Serve immediately, piping hot, with a ¼ piece of lime on the side.

Broiled Scallops

Pairings by dish star ingredient Dish star: clams

1. **Other stars: bacon,** chicken, lobster**, mussels**, pork (chops and sausage), **shrimp,** squid
2. **Allium or base ingredients: garlic, onions,** and shallots
3. **Vegetables:** carrots, **celery,** corn, green onions, leek, peas, spinach, sweet peppers
4. **Complex carb ingredients: potato,** rice**, pasta**
5. **Cruciferous vegetables:**
6. **Herbs:** sweet (basil, **parsley**) savory (**oregano, thyme**)
7. **Spices:** fresh (anise) earthy (bay leaf, **saffron**)
8. **Umami enhancers:** anchovy, **nori,** Worcestershire sauce
9. **Hot peppers (capsaicin):** cayenne, hot peppers (ají amarillo, red, green), paprika
10. **Condiments:** kosher salt, sea salt, lemon, lemon juice
11. **Other ingredients:** bread, breadcrumbs, white wine
12. **Fruits: tomato** and paste
13. **Dairy and fats: butter**, cream, milk, **olive oil,** cheese (Parmesan)
14. **Nuts and seeds:** fennel seeds.
15. **Simmering liquids:** chicken broth, **clam juice,** fish stock

Recipe example

Clams a la Mar

The clams, in this recipe, pair well with spinach, basil, ají amarillo, Nori and Dashi, lemon or lime juice, butter.

Servings: 4

INGREDIENTS

Umami bomb alternate

- 6 sheets of soaked nori in ½ cup dashi
- 1 cup baby spinach
- Tbsp. lemon or lime juice
- ¼ cup of cubed red onion, small.
- 1 cup basil leaves
- 1 stick unsalted butter
- ají amarillo to taste.
- One Tbsp. chopped garlic.
- 16 Clams on the half shell.

INSTRUCTIONS

1. In a hot pan with moderate heat, sauté garlic, onions, ají amarillo until caramelized; add soaked nori with dashi to deglaze. Then add spinach and basil and cook until soft. Squeeze excess water and while hot, puree mixture in food processor. Add soft butter and mix well.

2. Top each clam with the mixture and broil in max broil for 2-3 minutes and serve immediately.

Clams a la Mar.

Pairings by dish star ingredient Dish star: **oysters**

1. **Other stars:** bacon, caviar, crabmeat, eggs, sausage (pork), shrimp
2. **Allium or base ingredients: garlic, onions**, shallots
3. **Vegetables:** artichokes, **celery**, green onions, green and other sweet peppers, mushrooms, okra, spinach
4. **Complex carb ingredients: potatoes**, risotto
5. **Cruciferous vegetables:** none
6. **Herbs:** sweet (basil, **parsley**) savory (thyme) - **avoid** anise and tarragon
7. **Spices:** earthy (bay leaf)
8. **Umami enhancers: nori, bonito flakes**, Worcestershire sauce
9. **Hot peppers (capsaicin):** cayenne, cajun seasoning, hot sauce, peppers (ají amarillo, red), paprika
10. **Condiments:** lemon juice, salt (kosher, sea)
11. **Other ingredients: breadcrumbs**, corn bread, corn meal, pernod, anisette, sherry, white wine
12. **Fruits: tomato, olives**
13. **Dairy and fats:** butter, cream (heavy), milk, olive oil, Parmesan cheese
14. **Nuts and seeds:** none.
15. **Simmering liquids:** none

Recipe example:

Oysters a la Mar

Use same recipe as for Clams a la Mar, replacing the clams with oysters; works best with meaty oysters, like Chesapeake Bay oysters.

Pairings by dish star ingredient Dish star: **tomatoes**

1. **Other stars:** anchovies, avocado, bacon, chicken, crab meat, grilled fish, ground beef, shellfish, shrimp
2. **Allium or base ingredients: garlic, onions**
3. **Vegetables:** carrots, celery, corn, cucumber, **green onion, green and other sweet peppers**, lettuce,
4. **Complex carb ingredients:** rice, **pasta**
5. **Tomato paste**
6. **Herbs:** sweet (**basil,** parsley) savory (oregano, thyme)
7. **Spices: sweet** (marjoram) earthy (cumin)
8. **Umami enhancers: anchovy**
9. **Hot peppers (capsaicin):** cayenne, hot peppers (ají amarillo, red, green)
10. **Condiments:** lemon juice, lime juice, vinegar, salad dressing, sugar, salt, and pepper
11. **Other ingredients:** breadcrumbs, red wine, sugar
12. **Fruits:** avocado, watermelon
13. **Dairy and fats: butter**, cheese (feta, cheddar, mozzarella, ricotta, **Parmesan**), mayonnaise, **olive oil**
14. **Nuts and seeds:** none.
15. **Simmering liquids:** chicken broth
16. **Classic flavors pairings:**

Tomato + avocado + basil (+ crab meat + lemon)

Tomato + basil + goat cheese + balsamic vinegar + EVOO

Tomato with watermelon salad - José Andrés

Tomato + mayonnaise + strattu[16] (tomato paste) **with squid** (Eggleton, 2015)

Recipe example

Slow Roasted Tomatoes

This is a side dish that is easy to prepare and can be utilized in truly diverse dishes. It is therefore a base recipe. Slow, prolonged cooking allows the tomato to dehydrate and to conserve the fragrance and consistency and intensify the flavor.

In this recipe the tomatoes pair well with garlic, thyme, EVOO, confectioner's sugar, salt, and black pepper. *Seafood Paella, Larger pan*

INGREDIENTS
2 lb. red tomatoes
1 garlic clove
2 sprigs of thyme
EVOO
1 tsp. of confectioners' sugar
Salt to taste

INSTRUCTIONS
1. Cut the tomato in half and place on a cookie sheet covered with parchment paper.
2. Condiment the tomatoes with salt, thyme, olive oil, and dust with confectioners' sugar.

---

[16] Italy magazine, Sicily's Signature Tomato Paste: 'Strattu, accessed date June 23, 2021,

https://www.italymagazine.com/news/sicilys-signature-tomato-paste-strattu#:~:text='Strattu%20%5Bthe%20dialect%20word%20for,up%20pasta%20sauces%20no%20end

3. Bake for 150 minutes at 320°F, remove from heat and let them cool.

If in a rush you could Stove top roast canned San Marzano tomatoes with their juices at a low temperature simmering until the consistency of a purée is achieved.

Recipe example

Blini with white anchovies and Slow Roasted tomatoes

Inspired by Chef Thomas Keller

In this recipe, the tomatoes are paired well with chicken stock, parsley, salt, potatoes, butter, salt, and white anchovies.

Servings: 6

INGREDIENTS

¼ cup minced slow roasted tomatoes (about 6-8 pieces)
1½ Tbsp. vegetable or chicken stock
2 tsp. EVOO
3 Tbsp. Beurre Monté
1½ tsp. minced Italian parsley.
Kosher salt
12 Yukon Gold potato blinis
½ anchovy fillet per blini. I like to soak the anchovies in milk for about an hour, rinse them in cold water, pat them dry with paper towels and then soak them in fresh olive oil, extra virgin. Then you may put the oil in a small serving dish used to soak pieces of baguette bread.

INSTRUCTIONS

1. Warm the roasted tomatoes and chicken stock in a small saucepan.
2. Stir in the EVOO and simmer for a few seconds.
3. Reduce the heat and stir in the Beurre Monté, parsley, and salt to taste.
4. Place spoonful of sauce in each plate and top with 2 blini and place ½ an anchovy over each blini.

5. I often add to the anchovies a small dollop of caviar and Crème Fraiche.

Yukon Potatoes Blinis

This recipe was published by Bon Appétit[17] (see foot note) and credited as being Inspired by Chef Thomas Keller.

INGREDIENTS

1 pound Yukon Gold potatoes, scrubbed.
3 large egg yolks, room temperature
1 large egg, room temperature
½ cup sour cream, room temperature
¼ cup all-purpose flour
1½ teaspoon kosher salt
½ teaspoon finely ground black pepper
¼ teaspoon baking soda
¼ teaspoon ground nutmeg
2 tablespoons unsalted butter, melted, slightly cooled,
Special equipment: a food mill or potato ricer

INSTRUCTIONS

Preheat oven to 425°. Prick potatoes all over with a fork and bake on a rimmed baking sheet until very tender and a knife slides easily through flesh, 60-70 minutes. Let cool slightly. Reduce oven temperature to 200°.

Meanwhile, whisk egg yolks, egg, and sour cream in a medium bowl to combine; set sour cream mixture aside.

Cut potatoes in half lengthwise and scoop flesh from skins, discard skins. Pass flesh through food mill or ricer fitted with the small-hole disk into a large bowl. Working quickly, sprinkle flour, salt, pepper, baking soda, and nutmeg over potatoes; toss lightly with a fork to distribute ingredients, fluff potatoes, and break up any clumps. Make a well in the center and pour in reserved sour cream mixture. Whisk in a circular motion, working from the

---

[17] Bon Appetit, Potato Blini, accessed date June 23 2021,
https://www.bonappetit.com/recipe/potato-blinis

center out to incorporate, just until smooth (it should look like a thick pancake batter). Cover; let sit 10 minutes.

Blinis over Roasted Tomatoes with anchovies, caviar, and Crème Fraiche

Heat a large skillet, preferably nonstick or cast iron, over medium-low. Brush skillet with a thin layer of butter. Spoon scant tablespoon full of batter into skillet, spacing about 1" apart. Cook blinis until undersides are golden brown and surfaces look matte and bubbles form on top, about 90 seconds. Gently flip and cook until other sides are golden brown, about 1 minute. Transfer to a wire rack set inside a rimmed baking sheet and keep warm until ready to serve (up to 1 hour before). Repeat with remaining batter, wiping out skillet between batches and brushing with more butter.

Pairings by dish star ingredient Dish star: **beans**

1. **Other stars: bacon**, beef, chicken, pork (sausage, chops), **ground beef**
2. **Allium or base ingredients: garlic, onions**
3. **Vegetables: carrots, celery**, corn, green onions, **green sweet pepper and others,** lettuce, mushrooms
4. **Complex carb ingredients: kidney beans, pinto beans**, potato, rice, pasta
5. **Cruciferous vegetables:** none

6. **Herbs:** sweet (basil, cilantro, parsley) savory (oregano, thyme)
7. **Spices:** earthy (cumin, bay leaf)
8. **Umami enhancers:** mustard, soy sauce
9. **Hot peppers (capsaicin):** cayenne, **chili powder**, peppers **(ají amarillo, red, green)**
10. **Condiments:** salt, sugar, molasses, brown sugar
11. **Other ingredients:** tortilla chips
12. **Fruits: tomato**, ketchup
13. **Dairy and fats: butter, cheese (cheddar), EVOO**
14. **Nuts and seeds:** none.
15. **Simmering liquids:** chicken broth, tomato sauce

How to cook dried beans

Place **1 lb. dried pinto beans or another kind** in a large heavy pot. Cover with water about 2" above top of beans. Cover pot, bring to a boil, and then remove from heat. Let rest 1 hour. Stir in **1-1/2 tsp. kosher salt** and bring to a boil over medium heat. Uncover, reduce heat, and simmer until beans are tender and creamy, checking after 1 hour and adding more water as necessary to keep beans submerged.

Recipe example

Escabeche of beans

In this recipe, the beans pair well with eggs, garlic, onions, bay leaf, cumin, ají panca paste, ají amarillo (both available at most Latin markets), olive oil, vinegar.

Servings: 6

INGREDIENTS

1 lb. Cannelli beans
Salt and pepper to taste
4 Tbsp. olive oil or vegetable oil

1 medium red onion in thick slices

3 garlic cloves pressed or minced.

1 bay leaf

1 Tbsp. ají panca paste.

2 ajíes amarillo in thin slices across or paste.

 8 Tbsp. red wine vinegar

1 tsp. ground cumin

4 Alonso olives

2 hard-boiled eggs

INSTRUCTIONS

1. Cook the beans as per the instructions on how to cook beans.

2. Prepare the soffrito (aderezo) for the escabeche sauce: In a hot skillet add the oil and when shimmering with the heat at medium-low, add the sliced onion, minced garlic, ají panca, ají amarillo, bay leaf, vinegar, cumin, salt, and pepper. Cook until onions begin to soften but are still crunchy (5-6 minutes).

3. Add the vegetable stock and cook for a few more minutes (taste it, if too spicy cook some more but keep the onions still crunchy).

Place the beans in a serving bowl, cover with the escabeche sauce, and decorate with the olives and hard-boiled eggs. May be accompanied by sautéed fish like grouper or another white fish.

Cannellini Beans Escabeche

Recipe example

Chili con carne and beans

In this recipe the beans pair well with beef, garlic, onion, green pepper, celery, tomatoes, capsicum (red pepper), and salt.

Servings: 4 1.5-cup servings

INGREDIENTS

For soffritto:

1 cup chopped onion.
2 Tbsp. of ají amarillo paste or chili of your choice.
1 Tbsp. ground cumin
1 tsp. bottled minced garlic.
1/4 tsp. ground red pepper
1-pound ground sirloin

For sauce:

1 ½ cups of crashed tomatoes concasse or canned Marzano.
1/2 tsp. salt
1 (14.5 oz.) can diced tomatoes with green peppers, celery, and onion.1- 19 oz. canned kidney beans, rinsed and drained, or 1 lb. dry kidney beans.

INSTRUCTIONS

1. Cook the first 6 ingredients in a large nonstick skillet over medium-high heat until the beef is browned, stirring to crumble. Add crushed tomatoes, salt, and diced tomatoes; bring to a boil.

2. Cook 4 minutes, stirring occasionally. Add cooked or canned beans; cook 2 minutes or until thoroughly heated.

Pairings by dish star ingredient Dish star: **lentils**

Other stars: barley, pork (sausage, bacon)

1. **Allium or base ingredients: garlic,** ginger, **onions**
2. **Vegetables: carrots, celery**, green onions, green sweet peppers and others, leeks, mushrooms, peas, split peas, spinach
3. **Complex carb ingredients:** potatoes, **rice, pasta**
4. **Cruciferous vegetables:** none
5. **Herbs:** sweet (basil, cilantro, **parsley**) savory (oregano, **thyme**)
6. **Spices:** sweet (cinnamon, cloves) earthy (**bay leaf, cumin**, turmeric)
7. **Umami enhancer:** none
8. **Hot peppers (capsaicin):** cayenne, curry, hot peppers (amarillo, red, green)
9. **Condiments:** lemon, lemon juice, balsamic vinegar, **red wine vinegar,** salt, black pepper
10. **Other ingredients:** none
11. **Fruits: tomatoes,** tomato paste
12. **Dairy and fats: butter, EVOO,** yogurt
13. **Nuts and seeds:** none.
14. **Simmering liquids:** stocks (beef, **chicken, vegetables**)

Recipe example

Lentil salad with creole Peruvian onion sauce

Lentils are high in fiber, soluble (pectin), and complex carbohydrates, while low in fats and calories. Their high protein content makes lentils a perfect option for those looking to boost their protein intake. They are naturally gluten free, ideal in a gluten free kitchen.

In this recipe, the lentils pair well with garlic, onion, bay leaf, thyme, lemon juice, olive oil, salt, and black pepper.

Servings: 4½ cups

INGREDIENTS

1 cup dry brown lentils

1 bay leaf

2 sprigs fresh thyme

½ medium to large finely diced red onion

1 clove garlic, minced.

2 Tbsp. lemon juice

3 Tbsp. olive oil

One tsp. kosher salt

Two tsp. ají amarillo paste.

Fresh ground black pepper

INSTRUCTIONS

1. In a medium saucepan combine lentils, bay leaf, and thyme.
2. Add enough water to cover by 1 inch.
3. Bring to boil, reduce heat and simmer uncovered until lentils are tender but not mushy, about 16 to 20 minutes.

4. Drain lentils and discard bay leaf. Let the lentils cool.

5. Wash the raw onions by first mixing them with 2 Tbsp. of salt, massage them with your hands for a minute or so, and then rinse with hot water.

6. Place in a large bowl: red onion, parsley, garlic, lemon juice, olive oil, salt, and pepper.

7. Toss to combine and serve chilled or at room temperature.

Lentil salad with creole onion sauce and sweet piquant bell peppers

Pairings by dish star ingredient Dish star: Quinoa

1. **Other stars:** barley, **black beans**, chicken, chickpeas, eggs, guar gum, tofu, wheat.

2. **Allium or base ingredients: garlic,** ginger, **onions,** shallots

3. **Vegetables: carrots, celery, corn, green onions, green pepper and other sweet peppers, mushrooms,** peas, salad greens, spinach, tofu, yellow squash, zucchini.

4. **Complex carb ingredients:** potatoes, rice, tapioca starch

5. **Cruciferous vegetables: cabbage**

6. **Herbs:** sweet (basil, **cilantro, mint, parsley**) savory (bay leaf, oregano, thyme) earthy (**cumin**)

7. **Umami enhancers:** soy sauce

8. **Spices: cardamom, vanilla, smoked bacon, mint, ginger, thyme**

9. **Hot peppers (capsaicin):** cayenne, hot chilies (**ají amarillo, red, green**), curry

10. **Condiments:** citrus (**lemon juice, lime juice,** lemon and lime zest, orange, orange juice), salad dressing, salt (kosher, **sea salt**), vinegar (balsamic, rice)
11. **Other ingredients:** sugar, honey, white wine
12. **Fruits:** apples, cranberries, currants, dried fruits, olives, orange, raisins, **tomatoes)**
13. **Dairy and fats: butter,** feta cheese, milk, **EVOO**
14. **Nuts and seeds: almonds**, flax seeds, pecans, pine nuts, sunflower seeds, walnuts.
15. **Simmering liquids:** stock **(chicken, vegetable)**

Recipe example

Quinoa soufflé

In this recipe the quinoa pairs well with eggs, milk, cream, butter, Parmesan cheese, nutmeg, thyme, black pepper.

INGREDIENTS

7 oz. quinoa
White sauce ingredients:
3 Tbsp. flour
3 Tbsp. butter
1/2 tsp. nutmeg
1/4 tsp. black pepper
1/4 tsp. dried thyme
8 oz. evaporated milk
6 eggs
¼ pint cream
5 oz. Parmesan cheese
1 tsp. cream of tartar

INSTRUCTIONS

1. Cook the quinoa in water that covers it by an inch for 20 minutes. Proportion of 1 part of quinoa for 1 ½ parts of water by volume.
2. Make a Béchamel sauce by sautéing flour in butter until bubbly and with nutty aroma; add seasonings and warm milk.
3. Beat the egg yolks and mix with the hot Bechamel after tempering (mixing a hot ingredient with an egg-containing mix and preventing the eggs from becoming scrambled eggs), mix a bit of the hot ingredient with the egg mix first, so the temperature does not rise too fast, then add a bit more of the hot ingredient, and the egg mix will be tempered and you can then mix it all up); put everything together, and add the cream and cheese (3½ oz.).
4. Beat the egg whites until stiff.
5. Fold the quinoa and the egg yolk preparation into the egg whites.
6. Place in a refractory mold (preferred, a mold that will tolerate the heat of baking, e.g., Pyrex mold or a soufflé mold), well-greased and floured. Sprinkle with remaining grated cheese. Bake at 350° F for 1/2 hour, until surface nicely browned.

Pairings by dish star ingredient Dish star: cauliflower

1. **Allium or base ingredients: garlic,** ginger, **onions**
2. **Vegetables: carrots, celery,** green onions, green sweet peppers and others, kale lettuce, mushrooms, peas
3. **Complex carb ingredients: potatoes, pasta**
4. **Cruciferous vegetables: broccoli**
5. **Herbs:** sweet (basil, parsley)
6. **Spices:** fresh (coriander seeds) sweet (nutmeg) earthy (cumin, turmeric), garam masala
7. **Umami enhancer:** mustard
8. **Hot peppers (capsaicin):** cayenne, hot peppers (ají amarillo, red, green), curry, paprika
9. **Condiments:** lemon juice
10. **Other ingredients:** breadcrumbs, sugar
11. **Fruits:** olives, tomato

12. **Dairy and fats:** butter, cheese (cheddar, soft cheeses), mayonnaise, **milk,** sour cream, **olive oil**, salad dressings
13. **Nuts and seeds: cashews.**
14. **Simmering liquids:** stocks (chicken, vegetable)

Recipe example

Cauliflower based mock Alfredo sauce

In this recipe, cauliflower pairs well with garlic, onion, breadcrumbs, lemon zest, lemon juice, parsley, white miso, salt.

INGREDIENTS

Cashews raw 1 cup
Cauliflower florets 4 cups
Fettuccini 8 oz. or konjac fettuccini (like It's skinny®)
Breadcrumbs ½ cup ?
Parsley 1 Tbsp. minced.
Lemon zest ½ tsp.
Kale julienne 4 cups
Lemon juice 1 Tbsp.
White miso 2 Tbsp.
Garlic, onions (cubed small)
Salt and water 1 cup

INSTRUCTIONS

1. Boil water and add the cauliflower and cashews. Boil for about 15 minutes and then reserve.
2. Add noodles to the boiling water, cook the noodles 2 minutes shorter than to al dente and add the kale, boil for another 2 minutes and drain. Keep them in the pot, but not in the water (if low carb diet is desired use Konjac fettuccini).
3. Add all other ingredients to a blender (but not the breadcrumbs nor lemon zest) and puree to a sauce consistency. If necessary, add nonfat evaporated milk or water depending on desired sauce consistency.

4. Place the noodles and kale in a serving dish, add the sauce, and mix. Finish with the breadcrumbs and lemon zest mix. May broil it for a minute or 2 and serve immediately.

Fettuccini with Mock Alfredo sauce

Pairings by dish star ingredient Dish star: broccoli

1. **Other stars:** bacon, **cauliflower, chicken,** eggs, ham
2. **Allium or base ingredients: garlic**, ginger, **onion,** shallot
3. **Vegetables**: **carrots, celery, green onions**, green sweet pepper and others, mushrooms, peas, and zucchini
4. **Complex carb ingredients:** potato, rice, pasta
5. **Cruciferous vegetables:** cauliflower
6. **Herbs:** sweet (basil)
7. **Spices:** none
8. **Umami enhancers:** soy sauce
9. **Hot peppers (capsaicin):** hot peppers (red)
10. **Condiments:** lemon juice, white vinegar, salt
11. **Other ingredients:** sugar
12. **Fruits:** raisins, tomato
13. **Dairy and fats**: **butter,** cheese (**cheddar**, Parmesan), **mayonnaise, milk, olive oil,** salad dressings
14. **Nuts and seeds:** sunflower seeds.
15. **Simmering liquids:** stocks (chicken, vegetable)

Recipe example

Blanched Broccoli

INSTRUCTIONS

How to boil broccoli to perfection – according to the Thomas Keller method[18] (Keller, 1999)

This method protects the appearance of the vegetable and retains texture and flavor.

Big pot blanching: Blanching green vegetables should be done in a big pot with a lot of water and a lot of salt (1/4 cup for a 4-quart pot of water) until they are fully cooked for serving conditions. It is a color issue, and vegetables should be bright green: "We taste first with our eyes." Green vegetables raw or improperly cooked appear dull because a layer of gas develops between the skin and the pigment. Fast and hot cooking prevents this.

Using a large pot and cooking with lots of water accomplishes the following:

- It avoids dropping the temperature of the liquid as you add the vegetables because the proportion of water to vegetables prevents it. Also, at the boiling point the acids and enzymes released from the vegetables are destroyed.
- The salt prevents color from leaching into the water. On addition, a plus, is that the veggies will be properly seasoned when they are done (the water should taste like the ocean).
- Avoid cooking the vegetables excessively, by plunging into a large quantity of ice water immediately after removal from the boiling water. Leave them until they are chilled through, then drain them. Store in a dry container up to a day until ready to use.

---

[18] MasterClass.com, "How to Cook Broccoli: How to Steam Broccoli and Roasted Broccoli Recipe" 12/08/2021, https://www.masterclass.com/articles/how-to-cook-broccoli-how-to-steam-broccoli-and-roasted-broccoli-recipe

Pairings by dish star ingredient Dish star: Brussels sprouts

1. **Other stars: bacon,** broccoli, **bean sprouts,** chicken, eggs
2. **Allium or base ingredients: garlic, onions,** shallots
3. **Vegetables:** carrots, green onions
4. **Complex carb ingredients:** pasta
5. **Cruciferous vegetables:** none
6. **Herbs:** sweet (dill, parsley) earthy (thyme)
7. **Spices:** cardamom, cumin, curry, ginger
8. **Umami enhancers:** Dijon mustard, soy sauce
9. **Hot peppers (capsaicin):** none
10. **Seasonings/ Flavorings:** kosher salt, balsamic vinegar, cider vinegar, lemon juice, sugar
11. **Other ingredients:** breadcrumbs
12. **Fruits:** none
13. **Dairy and fats: butter,** cheese (Parmesan), heavy cream, milk, **olive oil**
14. **Nuts and seeds:** chestnuts, pecans, walnuts.
15. **Simmering liquids: chicken broth**

Recipe example

Brussels' sprout hash with caramelized shallots

In this recipe the Brussels sprouts pair well with chicken, shallots, salt, cider vinegar, sugar, EVOO, water.

INGREDIENTS

6 tablespoons (3/4 stick) butter, divided.
½ pound shallots, thinly sliced
Coarse kosher salt
2 tablespoons apple cider vinegar
4 teaspoons sugar or stevia granulated sweetener.
1 1/2 pounds Brussels sprouts, trimmed.
3 tablespoons extra virgin olive oil
1 cup water

INSTRUCTIONS

1. Melt 3 tablespoons butter in medium skillet over medium heat. Add shallots, sprinkle with coarse kosher salt and pepper. Sauté until soft and golden, about 10 minutes. Add vinegar and sugar. Stir until brown and glazed, about 3 minutes.

2. Halve Brussels sprouts lengthwise. Cut lengthwise into thin (1/8 inch) slices. Heat oil in large skillet over medium-high heat. Add sprouts, sprinkle with salt and pepper. Sauté until brown at edges, 6 minutes. Add 1 cup water and 3 tablespoons butter. Sauté until most of the water evaporates and sprouts are tender but still bright green, 3 minutes. Add shallots, season with salt and pepper. Serve over sautéed chicken.

Recipe example

Sautéed Brussels' sprouts

This recipe takes advantage of the Brussels sprouts pairing with butter, garlic, salt and pepper, and lemon juice.

Ingredients

Brussels sprouts, butter, garlic, salt, pepper, and lemon for the lemon juice; in amounts and proportions appropriate for the number of servings needed.

INSTRUCTIONS

1. In a large skillet over medium-high heat, heat oil or butter. Once hot, add Brussels sprouts and shake pan so all cut sides of sprouts settle cut side down in a single layer. Cook undisturbed 5 to 7 minutes, until the cut sides of sprouts become caramelized.

2. Stir and continue to cook until sprouts have taken on color and become tender, 6 to 8 minutes more. Season with salt and pepper, squeeze lemon over, and add garlic. Stir and cook 1 minute longer.

3. Spread the bottom of the serving plate with balsamic Glaze. The sweets of the glaze enhance the flavor of the sprouts.

Pairings by dish star ingredient Dish star: cabbage

1. **Other stars: bacon**, beef, **chicken**, ground beef, pork chops, sausage
2. **Allium or base ingredients: garlic**, ginger, **onions**
3. **Vegetables: carrots, celery, green onions, sweet peppers**
4. **Complex carb ingredients: potato**, rice, **pasta**
5. **Cruciferous vegetables:** none
6. **Herbs:** sweet (basil)
7. **Spices:** sweet (caraway seeds, anise seeds)
8. **Umami enhancer:** soy sauce
9. **Hot peppers (capsaicin):** hot peppers (amarillo, red, green)
10. **Condiments:** lemon juice, vinegar (cider, rice, white)
11. **Other ingredients:** none
12. **Fruits:** apple, tomato
13. **Dairy and fats: butter,** mayonnaise, milk, olive oil
14. **Nuts and seeds:** celery and sesame seeds.
15. **Simmering liquids:** stocks (beef, chicken)

Recipe example

Fried cabbage with bacon

This recipe takes advantage of the pairings of cabbage with bacon, chicken broth, garlic, onions, anise seeds, ají amarillo, rice vinegar, sugar.

INGREDIENTS

6 slices bacon, chopped.
1 onion, diced in small cubes.
2 cloves garlic, minced or chopped.
1 large head green or red cabbage, cored and sliced.
1 tsp. chicken bouillon, (or seasoned salt), may dissolve in 2 Tbsp. of water.
Salt and cracked black pepper, to taste.
1/4 tsp. Cajun seasoning, or Old Bay or smoky paprika (pimentón).
2 tomatoes concasse and dried in oven at 250°F for 2 hours.

INSTRUCTIONS

1. To prepare the tomatoes, core the stem attachment with a paring knife and cut the skin in a cross at the opposite end. Have a pot of boiling water ready, place the tomatoes in it for a couple of minutes, remove and cool them in an ice bath. Peel the tomatoes, starting from the crosscut. Then cut the tomatoes across (not lengthwise) and remove the seeds and juice with a small teaspoon. At this point, cut them in half again, stretch them on a cookie sheet and bake at 250°F or 120°C for a couple of hours.

2. Cook bacon in a large skillet over medium heat until crispy. Transfer bacon using a slotted spoon to a plate. Do not discard bacon drippings.

3. Sauté the onion in the bacon grease until the onion caramelizes (about 8 minutes). Add the garlic and cook until fragrant (30 seconds).

4. Stir in the cabbage and cook for a further 8 minutes while stirring occasionally. Season with the bouillon, salt, pepper, and Cajun seasoning (or paprika).

5. Mix bacon into the cabbage to warm through and serve.

Recipe example

Cabbage wrapped ground pork meat rolls

INGREDIENTS

Servings: About 24 rolls, 8-12 servings

1/2 cup milk, 1/2 cup fine breadcrumbs or alternatively 1 cup cooked rice small grain boiled.
2 large eggs.
2 tsp. salt
Freshly ground black pepper
1/2 cup grated Parmesan cheese.
1/4 cup finely minced Italian parsley.
1pound ground meat (such as beef, pork, turkey, chicken, or veal, or a mix)
½ lb. shiitake mushrooms sliced
1/2 cup finely chopped onion (or grated on a Coarse grater) from 1 small yellow onion.
1Bunch scallions chopped whites only.
1clove garlic (finely minced)

1 Tbsp. of ají amarillo
1 inch ginger grated.
½ tsp. sesame seeds
2 large heads Napa Cabbage

INSTRUCTIONS

1. **For this recipe,** we will use ground pork meat, which takes advantage of good pairings of pork, with eggs, garlic, onion, cheese, and sesame seeds. Pork is also paired with parsley, ají amarillo, and breadcrumbs or rice. The ají amarillo serves as a bridge between the cabbage and pork since it pairs well with both.

2. **Prepare the Napa cabbage leaves**. Remove 24 outer leaves from both cabbage heads, 12 from each. Roll each leaf with a rolling pin gently to flatten and smooth it. If necessary to soften the cabbage leaves, dip them in boiling water for a few seconds and then roll them with the rolling pin.

3. Core the remaining cabbage and chop it very fine. Place them in a colander, sprinkle with salt and set aside.

4. **Prepare the bonding for the filling of pork.** Place the breadcrumbs or rice in a small bowl, pour in the milk, and stir to combine. Set aside while preparing the rest of the meatball mixture. The breadcrumbs will absorb the milk and become soggy.

5. **Prepare the flavorings for the filling of pork.** Whisk the egg in a large bowl until thoroughly mixed. Add the Parmesan, parsley, salt, a generous quantity of black pepper, and the ají amarillo paste. Whisk to combine.

6. **Add the protein.** Add the ground pork meat to the egg mixture. Use your fingers to thoroughly mix the egg mixture into the ground meat (don't knead it).

7. **Combine all the parts**. Add the onions, garlic, and soaked breadcrumbs. Mix them thoroughly into the meat with your fingers. Try not to overwork the meat; pinch the meat between your fingers rather than kneading it. Add the finely chopped cabbage in the colander, squeezing it to remove

the remaining water. Mix it with the pork mixture. Add the shiitake mushrooms.

8. **Prep the cabbage leaves**. Lay the leaves of cabbage one at a time with the stem end towards you. Put about 2/3 cup of the filling mixture onto the stem end of the cabbage leaf. Fold the sides and roll up the leaf and place the bundle seam side down in a baking dish.

9. **Repeat the process with all the remaining prepared cabbage leaves.** May need two baking dishes. If any filling is left, make them into balls and place them between the cabbage's rolls.

10. **Make a sauce in a small bowl** mix 1/4 cup of rice vinegar, 1 cup unsalted chicken broth, 1 tbsp. roasted sesame oil, 1 tsp. sugar, 1 Tbsp. ají amarillo paste. Pour over the cabbage rolls. At this point you may cover them and refrigerate for up to 24 hrs. or be frozen for later use

11. **Roast or broil the rolls in the oven uncovered.** Roast at 400°F for 25 to 30 minutes.

12. **(Watch closely if making rolls with lean meat.)** The outsides may be browned, and they register 160°F in the middle on an instant-read thermometer.

13. **Encrust** a few sesame seeds in each roll.

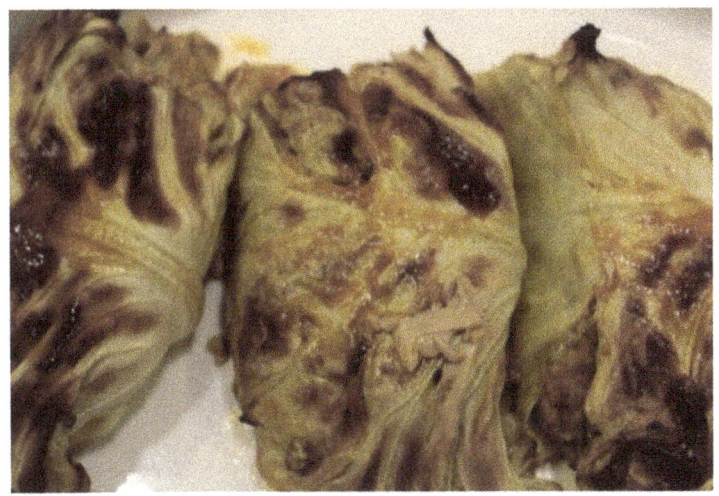

Cabbage Rolls

Chapter 7

Dishes from Your Pantry

"One of the very best things about life is the way we must regularly stop, whatever it is we are doing and devote our attention to eating."
—Luciano Pavarotti (Pavarotti, 2021)

Often or even usually, we find ourselves unable to run to the grocery store and we must use our creativity and create something delicious from what we already have on hand—what's in our pantry and refrigerator now.

One thing you can do in a case like that is to type into google "Recipe containing (as an example) chicken carrots cauliflower potatoes" and see what you get.

One result I got when doing it was:

Sheet pans chicken and vegetables by Delightful Mom Food, Healthy Gluten Free Recipes.

Sheet pan chicken and vegetables is one easy, balanced, gluten free, grain free meal ready in 30 minutes with only a pan and bowl to wash! Honey, lemon, and herbs soaked around a juicy chicken, then baked with carrots, broccoli, cauliflower, and potatoes is so tasty, healthy, and fits a Low Carb, Paleo, and gluten free diet!

For the recipe, visit the website.

I hope these ideas will make it easier for you to create your own dishes and to use recipes as memory crutches to make them reproducible or to stimulate your creativity when you find yourself in a rut.

Purposely, some of the recipes are somewhat vague, particularly the ones like sauces, or simmering, stewing, and casseroles, to stimulate your independence from recipes and help you rely more on cooking methods and give free rein to your creativity. This I believe can be achieved by having a better understanding of the science that underlies all cooking methods.

When creating your own recipes from what is available in your refrigerator and/or pantry follow these steps, they will facilitate your task.

- Select first what your star ingredient will be a protein (beef, poultry, fish or crustaceans, or beans and vegetables).

- Select from the pairings list for your star ingredient your extender like potatoes, rice, or mixed vegetables.

- Select from the pairing list for your star ingredient and your extender what flavorings or herbs and spices to use.

- Make sure the dish is balanced by having at least 3 of the tastes and flavors from the diagram provided at the end of Chapter 1, either complementing or balancing each other.

- Taste the dish as it progresses, add salt pepper or other flavorings to layer the flavors in the dish, and to check for degree of doneness.

Go and have fun.

Index

208

Bibliography

Milford, H. S. (1913). *The Complete Poetical works.* London: Oxford University Place.

Alfaro, D. (2019, 10 28). *The Spruce eats.* Retrieved 06 23, 2021, from https://www.thespruceeats.com: https://www.thespruceeats.com/all-about-simmering-995786

Baldwin, D. E. (2013). *Sous vide for the home cook 2nd Edition.* Incline Village, NV: Paradox Press.

Brantley, A. (2018, 5 29). *The Science of Taste: Why Some Foods Taste Good Together.* Retrieved from https://bcbstwelltuned.com: https://bcbstwelltuned.com/2018/05/29/the-science-of-taste-why-some-foods-taste-good-together/

Briscione, J., & Parkhurst, B. (2018). *Flavor Matrix, The Art and Science of Pairing Common Ingredients to create extraordinary Dishes.* Bosto, Newyork: Houghton Mifflin Harcourt.

Child, J. (2018, September 6). *No one is born a great cook.* (I. Food and Drink, Editor) Retrieved August 8/19/2021, 2021, from AthenaQuotes.com: https://athenaquotes.com/no-one-is-born-a-great-cook-one-learns-by-doing-j

Cloake, F. (2011, 8 18). How to Cook the Perfect Paella. *The Guardian.* Retrieved 6 23, 2021, from https://www.theguardian.com/lifeandstyle

Cowper, W. (1913). *The Complete Poetical Works.* (M. H. S., Ed.) London: Oxford University Press.

Delhindra. (2019, 9 05). *Chef Q!* Retrieved 8 23, 2021, from https://chefqtrainer.blogspot.com/2019/09/5: https://chefqtrainer.blogspot.com/2019/09/5-basic-recipes-of-french-mother-sauces.html

Dunning, D. (2017, April 25). *https://sciencing.com/chemical-senses-8464326.html*. Retrieved from sciencing.com: sciencing.com/chemical-senses

Eggleton, P. (2015, 02 05). *Sicily's Signature Tomato Paste*. Retrieved 08 25, 2021, from https:www.italymagazine.com: https://www.italymagazine.com/featured-story/sicilys-signature-tomato-paste-strattu#:~:text='Strattu%20%5Bthe%20dialect%20word%20for,up%20pasta%20sauces%20no%20end.

Fields, W. (n.d.). *BrainyQuotes.com/quotes/w c fields 108730*. (Brainy Quotes) Retrieved 8 14, 2021, from FieldsQuotes.com.

Garten, I. (2021, 08 19). *InaGartenQuotes BrainyQuotes.com*. (H. Thoreau, Z. Ziglar, Editors, & BrainyMedia Inc.) Retrieved 2021, from BrainyQuotes.com.

Hale, K. (1966). *Irrepresible Churchill: A Treasury of Wiston Churchill's wit.* Cleveland and New York: World Publishing Company.

Holmberg, M. (2012). *Modern Sauces.* San Francisco: Chronicle Books.

https://inspire.foodpairing.com/account/login. (n.d.). Retrieved from inspire.foodpairing.com: https://inspire.foodpairing.com/Account/Login?ReturnUrl=%2f&AspxAutoDetectCookieSupport=1

Keller, T. (1999). Big Pot Blanching. In T. Keller, *The French Laundry* (p. Location 1089). New York, New York, USA: Artisan/ Workman Publishing Inc. Retrieved from https://www.masterclass.com/how-to-cook-broccoli: , https://www.masterclass.com/articles/how-to-cook-broccoli-how-to-steam-broccoli-and-roasted-broccoli-recipe

Luis, A. (2021). *Table 1: Smoking Temperature of fats.* Stevensville MD.

Luis, A. (2021). *Table 2: Recommended Internal Temperature for Proteins.* Stevensville MD.

Luis, A. (2021). *Table 3: Heat Transfer According to its Nature.* Stevensville MD.

Luis, A. (2021). *Table 4: Sous Vide recommended Temps of Water Baths for degree of cooking.* Stevensville MD.

Luis, A. (2021). *Table 5: Sous Vide recommended time versus Thickness of the Product.* Stevensville MD.

Luis, A. (2021). *Table 6: Mother Sauces.* Stevensville MD.

Luis, A. (2021). *Table 7: Derivatives of the Mother Sauces.* Stevensville MD.

MayoClinic, S. (2021, 4 21). *Mayo Clinic.org/healthy-lifestyle/nutrition-and-healthy-eating/in-depth/fat/art-2004550.* Retrieved from www.MayoClinic.org: https://www.mayoclinic.org/healthy-lifestyle/nutrition-and-healthy-eating/in-depth/fat/art-20045550

Melis, M., & Barbarossa, I. T. (2017, June 9). Taste Perception of sweet, sour, salty, bitter, and umami and changes due to L-Arginine Suplementation. *Nutrients*, 541.

Meredith, L. (2016). *Air Fry Everything.* Philadelphia: Walah. Retrieved June 23, 2021, from Bluejean Cheff website/cooking school: https://bluej eanchef.com/cooking-school/air-fryer-cooking-charts/

Moncel, B. (2019, 9 19). *smoking-points-of-fats-and-oils.* Retrieved from https://www.thespruceeats.com: https://www.thespruceeats.com/smoking-points-of-fats-and-oils-1328753

Oz, F., & Kotan, G. (2016, 3 13). *https://doi.org/j.foodcont.* Retrieved from https://doi.org/10.1016/j.foodcont.2016.03.013

Page, K., & Dornenburg, A. (2008). *The Flavor Bible: The Essential Guide to Culinary creativity, Based on the Wisdom of America's Most Imaginative Chefs.* New York: Little, Brown & Company.

Patterson, D., & Mandy, A. (2017). *The Art of Flavour: Practices and Principles for creating Delicious Food.* New York, New York: Riverhead Books.

Pavarotti, L. (2021, 8 14). *https//www.goodreads,com/quotes/139486*. Retrieved from Goodreads.com.

Payal, M., & Ankita, R. M. (2018). *Mutagenicity: Assays and Applications.* Cambridge, MA: Academic Press/ Elsevier. Retrieved from https://www.sciencedirect.com/science/article/pii/B97801280925211 20012

Pub Chem estearic acid. (n.d.). Retrieved from ncbi.nim.nih.gov: https://pubchem.ncbi.nlm.nih.gov/compound/Stearic-acid

Pub Chem oleic acid. (n.d.). Retrieved from ncbi.nim.nih.gov: https://pubchem.ncbi.nlm.nih.gov/compound/oleic-acid

Roper, S. D., & Chaudhari, N. (2017). Taste Buds: cells, signals and synapses. *Nature revews Neuroscience, 18*, 485-497.

Saffitz, C. (n.d.). *https://www.brainyquote.com/quotes/claire_saffitz_1061636*. (S. L. Taylor, W. White, Editors, & BrainyQuote.com) Retrieved from BrainyQuote.com.

Saltus, E. (1917). *An Idler's Impression.* Chicago: Brothers of the Book page 20.

Saulnier, L. (1982). *Le Repertoire de La Cuisine.* Staines, Middlesex: Leon Jaeggi & Sons LTD.

Swati, B., & Santosh, J. (2016, 06 21). *https://doi.org10.1016/j.foodchem*. Retrieved from Journal Food Chem.: https://doi.org/10.1016/j.foodchem.2016.06.021

Tower, J. (1986). *New American Classics.* New York: Harper and Row Publishers Inc.

USDepartmentofHealthandHumanServices, & USDepartmentofAgriculture. (2015). *Dietary Guidelines for Americans 2015-2020.* Washington DC: http//health.gov/dietaryguidelines/2015/guidelines.

Wikipedia.org. (n.d.). Retrieved June 23, 2021, from
wikepidia.org/wiki/chinise cooking technics:
https://en.wikipedia.org/wiki/Edible_bird%27s_nest

Wikipedia.org/wiki/shark fin soup. (n.d.). Retrieved June 23, 2021, from
Wikepia.org/wiki/chinese cooking techniques:
https://en.wikipedia.org/wiki/Shark_fin_soup

www.ingramcontent.com/pod-product-compliance
Lightning Source LLC
Chambersburg PA
CBHW051147120626

46547CB00012B/974

9 781962 569477